Charles Lee Crandall

Tables for the Computation of Railway and other Earthwork

Charles Lee Crandall

Tables for the Computation of Railway and other Earthwork

ISBN/EAN: 9783744662239

Printed in Europe, USA, Canada, Australia, Japan

Cover: Foto ©berggeist007 / pixelio.de

More available books at **www.hansebooks.com**

FOR THE COMPUTATION

OF

Railway and Other Earthwork,

COMPUTED BY

C. L. CRANDALL, C. E.,

Associate Professor of Civil Engineering, Cornell University;
Member of American Society of Civil Engineers.

SECOND EDITION.

NEW YORK:
JOHN WILEY & SONS,
53 EAST TENTH STREET.
1893.

ANDRUS & CHURCH,
PRINTERS,
ITHACA, N. Y.

Preface to the First Edition.

THE object of these tables is to present a convenient aid in the Computation of Railway Earthwork. The volume is first found by the approximate method of averaging end areas, to which can be added, if desired, a correction due to the strict prismoidal formula, or a correction due to a modification of it for irregular ground. This separation reduces the labor, if both parts are computed at once, while it allows of computing only the approximate value at first, which will answer for all purposes except the final estimate, leaving the prismoidal correction to be computed at leisure on those volumes only where it is large enough to be appreciable.

The Table for "Volume by Averaging End Areas," is a table for triangular prisms, and it will give the approximate volume of triangular prismoids in cubic yards for a length of one hundred feet, between the parallel triangular ends, by taking the sum of the two numbers found with the base and altitude of each of the two end triangles as arguments. The table is thus independent of side slope, or width of roadbed, and will apply to borrow pits, trenches, etc., by dividing up the cross sections into triangles and trapezoids.

The method of dividing up the sections for three-level, five-level and irregular ground, while not essential, is believed to give the least labor in computing, either with or without tables.

The method for irregular ground is due to J. W. Davis, and was published in his *Formula for R. R. Earthwork*, 1876.

The Table for "Prismoidal Correction" is a table for triangular prismoids, and it will give the exact correction to be applied to the approximate volume found above for all solids which can be divided into triangular prismoids, having plane parallel ends, and planes or hyperbolic paraboloids for side faces. By using A. M. Wellington's method for irregular ground, that for the purposes of the prismoidal correction only, the ground may be called regular, the correction is easily and quickly applied, to three-level, five-level, and irregular ground. It may be noted that when either the bases or altitudes of the end triangles are equal, the correction disappears.

While quantities are only given for full feet the corrections for tenths are easily taken out from the margin without multiplication and can be added mentally with but slight increase of labor.

The Tables have been carefully recomputed in print, and it is confidently believed they are free from errors.

PREFACE TO THE SECOND EDITION.

The first edition was not electrotyped, and advantage has been taken of this to entirely recast the text, adding a proof of the prismoidal formula and a comparison with others in use, a formula for correction for curvature, and rules for cross sectioning.

ITHACA, N. Y., August, 1893.

Contents.

DERIVATION OF PRISMOIDAL FORMULA AND COMPARISON WITH OTHERS IN USE.

SEC.		PAGE.
1.	Prismoidal Formula	1
2.	Averaging End Areas	2
3.	Middle Areas	3
4.	Equivalent Mean Heights	3
5.	Triangular Prismoid with Curved Center Line	4

VOLUME FOR RAILWAY EARTHWORK IN CUBIC YARDS BY AVERAGING END AREAS.

6.	Formula for Three-level Ground	7
7.	Five-level Ground	7
8.	Irregular Ground	8
9.	Tables	10
10.	Example, Three level Ground	11
11.	Example, Five-level Ground	11
12.	Example, Irregular Ground	12
13.	Borrow Pits	13
14.	Sidehill Work	13
15.	End Sections	14
16.	Correction for Curvature	14
17.	Example, Correction for Curvature	16

PRISMOIDAL CORRECTION IN CUBIC YARDS.

18.	Three-level Ground	17
19.	Five-level Ground	17
20.	Irregular Ground, Borrow Pits, etc.	17
21.	Table of Prismoidal Corrections	18
22.	Example, Three-level Ground	18
23.	Example, Five-level Ground	19
24.	Example, Irregular Ground	19
25.	Example, Borrow Pits	20

COMPUTATION, END AREAS WITH PRISMOIDAL CORRECTION.

26.	Three-level Ground	21
27.	Irregular Ground	21
28.	Rules for cross-sectioning	23

Table of Triangular Prismoids for Railway Earthwork, in cubic yards, by Averaging End Areas.

Table of Prismoidal Corrections for Triangular Prismoids for Railway Earthwork.

Derivation of Prismoidal Formula and Comparison with Others in Use.

1. **Prismoidal Formula.**—In computing railway earthwork, it is customary to measure sections perpendicular to the center line at stations 100 feet apart, and at as many intermediate points as may be found necessary.

To find the volume between two such parallel sections :

Take at first the simple case shown in Fig. 1, where the section is straight from the center to the slope stake on each side, and where straight lines joining corresponding points of the upper lines of the sections lie in the ground surface. The ground surface thus formed is a hyperbolic parabo- loid, a characteristic property of which is that all planes parallel to the end sections cut the surface in straight lines, whose increments (or decrements) in length, starting from either end, are proportional to their distances from that end.

With the notation of Fig. 1, the area of the right-hand side of the first section (including the grade triangle under the roadbed),

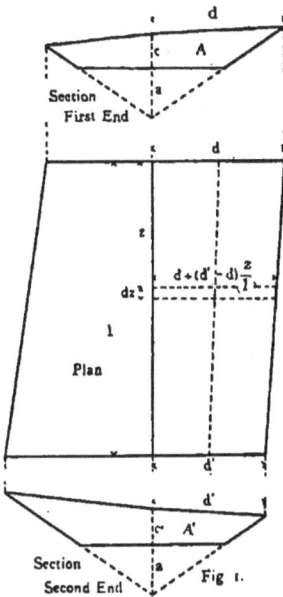

$$A = \tfrac{1}{2}(a+c)d.$$

At the distance z, since all the linear dimensions of the section can be interpolated,

$$A_z = \tfrac{1}{2}\left[a + c + (c'-c)\frac{z}{l}\right]\left[d + (d'-d)\frac{z}{l}\right] \tag{a}$$

Elementary volume of length dz = area into length

$$= A_z\, dz.$$

Integrating between the limits $z = 0$ to $z = l$,

$$\text{Volume} = \int_0^l A_z\, dz$$

$$= \tfrac{1}{2}\left[(a+c)dz+(a+c)(d'-d)\tfrac{z^2}{2l}+d(c'-c)\tfrac{z^2}{2l}+(c'-c)(d'-d)\tfrac{z^3}{3l^2}\right]^l_0$$

$$= \tfrac{1}{2}\left[(a+c)dl+(a+c)(d'-d)\tfrac{l}{2}+d(c'-c)\tfrac{l}{2}+(c'-c)(d'-d)\tfrac{l}{3}\right]$$

$$\text{Volume} = \frac{a+c}{2}\frac{dl}{6}+\frac{a+c'}{2}\frac{d'l}{6}$$

$$+\frac{a+c}{2}dl\frac{5}{6}+\frac{a+c}{2}\frac{d'l}{3}+\frac{c'-c}{2}\frac{d'l}{6}-\frac{a+c}{2}\frac{dl}{2}+\frac{c'-c}{2}\frac{dl}{6}\ .$$

$$= A\frac{l}{6}+A'\frac{l}{6}+\frac{a}{2}\frac{d+d'}{2}4\frac{l}{6}+\frac{c+c'}{4}\frac{d+d'}{2}4\frac{l}{6}, \text{ or}$$

$$\text{Vol.} = (A+A'+4A_m)\frac{l}{6} \tag{1}$$

where A_m is the area of the mid section found by making $z = l \div 2$ in (a).

i. e., *the true volume equals the sum of the areas of the end sections and four times that of the mid section, multiplied by one-sixth the perpendicular length between sections.*

This is the prismoidal formula, and the first proof of its application to an earthwork solid, as above, is due to Professor Gillespie.*

It is evident that an earthwork solid with polygonal end sections lying in parallel planes can be divided into solids with triangular bases as above, when there are the same number of sides in the end polygons and lines joining corresponding points of these sections lie in the bounding surfaces; therefore the prismoidal formula will give the true volume for all such solids. The solid thus described will be called a *prismoid*; *i. e.*, a prismoid will be defined as a solid having two plane polygonal parallel bases with the same number of sides in each, joined by either planes or hyperbolic paraboloids for side faces. With triangles for bases, the prismoid becomes a *triangular prismoid.*

2. **Averaging End Areas.**—This is a method in common use on account of its simplicity. Applying it to the triangular prismoid of § 1, and including the grade triangle for simplicity,

By averaging end areas,

$$\text{Vol.}_e = \frac{A+A'}{2}l$$

$$= \left[\frac{a+c}{2}d+\frac{a+c'}{2}d'\right]\frac{l}{2}.$$

True volume by the prismoidal formula,

$$\text{Vol.}_p = \frac{a+c}{2}d\frac{l}{6}+\frac{a+c'}{2}d'\frac{l}{6}+\frac{a}{2}\frac{d+d'}{2}\frac{2l}{3}+\frac{c+c'}{4}\frac{d+d'}{2}\frac{2l}{3}$$

Subtracting Vol.$_e$ from the true value, Vol.$_p$,

$$\text{Vol.}_p-\text{Vol.}_e = -\frac{a+c}{2}d\frac{l}{3}-\frac{a+c'}{2}d'\frac{l}{3}+\frac{d+d'}{2}a\frac{l}{3}+\frac{c+c'}{2}\frac{d+d'}{2}\frac{l}{3}$$

* See *Roads and Railroads*, New York, 1847.

Eq. 3]   *EQUIVALENT MEAN HEIGHTS.*   3

$$\text{Cor.}_e = (c - c')(d' - d)\frac{l}{12} \qquad (2)$$

where Cor._e is the correction to be applied to the volume found by averaging end areas in order to obtain the true volume.

Usually in passing from lighter to heavier work $c' > c$ and $d' > d$ (or conversely from heavier to lighter), making the above correction usually negative; *i. e.*, the method by averaging end areas gives on the average too large a volume.

3. **Middle Areas.**—The linear dimensions of the mid section will be the means of those of the end sections, giving

$$\text{Vol.}_m = \tfrac{1}{2}\left(a + \frac{c+c'}{2}\right)\frac{d+d'}{2}\,l$$

Subtracting this from the true value,

$$\text{Vol.}_p - \text{Vol.}_m = -\frac{a}{2}(d+d')\frac{l}{3} - \frac{cdl}{24} - \frac{c'd'l}{24} - \frac{cd'l}{8} - \frac{c'dl}{8} + \frac{a}{2}(d+d')\frac{l}{3} +$$
$$\frac{c+c'}{4}(d+d')\frac{l}{3}$$

$$\text{Cor.}_m = (c-c')(d-d')\frac{l}{24} \qquad (3)$$

where Cor._m is the correction to be applied to the volume found by middle areas in order to obtain the true volume.

This correction is one-half that found by averaging end areas, and has the opposite sign, *i. e.*, the method on the average gives results too small, but the error is only one-half as great as by the preceding method.

4. **Equivalent Mean Heights.**—In this method the center height of a level section having an equal area is found for each end section,

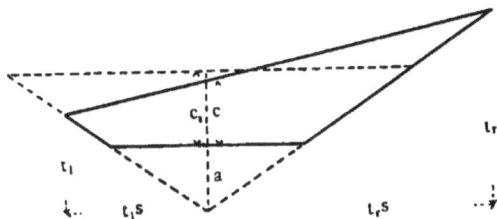

Fig. 2.

and the volume of the reduced solid is then accurately found by the prismoidal formula.

Applying this to two-level ground, with slope stake heights t_1, t_r, slope ratio s, reduced center height c_1, and distances out to slope stakes $t_1 s$ and $t_r s$,

$$A = \tfrac{1}{2}(t_1 + t_r)(t_1 + t_r)s - \tfrac{1}{2}t^2_1 s - \tfrac{1}{2}t^2_r s = t_1 t_r s$$

From the reduced area,

$$A = (a + c_1)^2 s$$

or,

$$a + c_1 = \sqrt{t_1 t_r}$$

The true volume of the *reduced* solid,

$$\text{Vol.}_{\text{e.m.h.}} = (A + A' + 4A'_m)\frac{l}{6}$$

$$= \left[(a+c_1)^2 + (a+c_1')^2 + 4\left(\frac{a+c_1}{2} + \frac{a+c_1'}{2}\right)^2 \right]\frac{sl}{6}$$

$$= \left[l_1 l_r + l_1' l_r' + (\sqrt{l_1 l_r} + \sqrt{l_1' l_r'})^2 \right]\frac{sl}{6}$$

The true volume of the original solid,

$$\text{Vol.}_p = (A + A' + 4 A_m)\frac{l}{6}$$

$$= \left[l_1 l_r + l_1' l_r' + 4\left(\frac{l_1 + l_1'}{2} \times \frac{l_r + l_r'}{2}\right) \right]\frac{sl}{6}$$

$$= [2l_1 l_r + 2l_1' l_r' + l_1 l_r' + l_1' l_r]\frac{sl}{6}$$

Subtracting the volume of the reduced solid from that of the original one,

$$\text{Vol.}_p - \text{Vol.}_{\text{e.m.h.}} = [l_1 l_r' - 2\sqrt{l_1 l_r \, l_1' l_r'} + l_1' l_r]\frac{sl}{6},$$

or, $$\text{Cor.}_{\text{e.m.h.}} = (\sqrt{l_1 l_r'} - \sqrt{l_1' l_r})^2 \frac{sl}{6} \qquad (4)$$

The correction is thus always positive, *i, e.*, the method always gives too small a result for two-level ground, except when $l_1 l_r' = l_1' l_r$; hence it will usually give too small results for ordinary ground.

5. **Triangular Prismoid with Curved Center Line.**—In curving the center line of the roadbed to the radius R, let the dimensions of

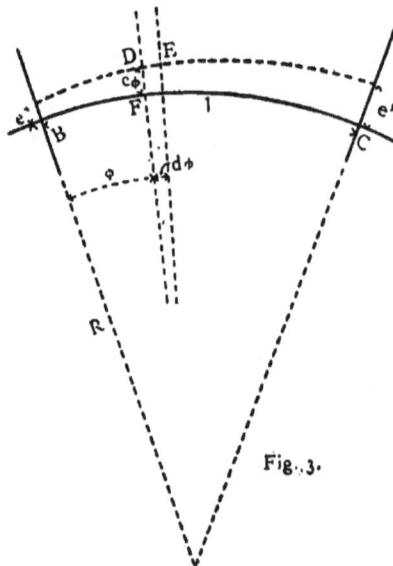

Fig. 3.

Eq. 4]   *CURVED CENTER LINE.*   5

all the cross sections remain unchanged. This will give the solid easiest to approximate in the field, and the one most like the prismoid, since all lineal dimensions of the cross section can be interpolated.

In Fig. 3, let BC be the center line of the roadbed; e, e_ϕ, e', the eccentricities of the sections, or the distances of their centers of gravity from the center of the roadbed; A, A_ϕ, A', the areas of the triangular sections, with bases b, b_ϕ, b', and altitudes a, a_ϕ, a'; ϕ the central angle from B to any point F; l the length BC of the solid along the center line of the roadbed, $= R\phi'$.

At any point D,

$$\text{Elementary volume} = A_\phi \,\overline{DE} = A_\phi\,(R + e_\phi)\,d\phi$$

$$\text{Volume} = \int_0^{\phi' = \frac{l}{R}} A_\phi\,(R + e_\phi)\,d\phi$$

Since all cross section dimensions can be interpolated,

$$A_\phi = \tfrac{1}{2}\Big[b + (b' - b)\frac{\phi R}{l}\Big]\Big[a + (a' - a)\frac{\phi R}{l}\Big]$$

$$e_\phi = e + (e' - e)\frac{\phi R}{l}$$

Substituting,

Volume =

$$\int_0^{\phi' = \frac{l}{R}} \tfrac{1}{2}\Big[b + (b' - b)\frac{\phi R}{l}\Big]\Big[a + (a' - a)\frac{\phi R}{l}\Big]\Big[R + e + (e' - e)\frac{\phi R}{l}\Big]d\phi$$

$$= \tfrac{1}{2}\Big[[ab\phi + a(b' - b)\frac{\phi^2 R}{2l} + b(a' - a)\frac{\phi^2 R}{2l} + (a' - a)(b' - b)\frac{\phi^3 R^2}{3l^2}]\,R$$

$$+ [ab\phi + a(b' - b)\frac{\phi^2 R}{2l} + b\,(a' - a)\frac{\phi^2 R}{2l} + (a' - a)\,(b' - b)\frac{\phi^3 R^2}{3l^2}]e$$

$$+ [ab\frac{\phi^2}{2} + a(b' - b)\frac{\phi^3 R}{3l} + b(a' - a)\frac{\phi^3 R}{3l} +$$

$$(a' - a)(b' - b)\frac{\phi^4 R^2}{4l^2}]\,(e' - e)\,\frac{R}{l}\Big]_0^{\phi' = \frac{l}{R}}$$

$$= \tfrac{1}{2}l\Big[ab + a(b' - b)\tfrac{1}{2} + b(a' - a)\tfrac{1}{2} + (a' - a)(b' - b)\tfrac{1}{3}$$

$$+ [ab + a(b' - b)\tfrac{1}{2} + b(a' - a)\tfrac{1}{2} + (a' - a)(b' - b)\tfrac{1}{3}]\frac{e}{R}$$

$$+ [ab\tfrac{1}{2} + a(b' - b)\tfrac{1}{3} + b(a' - a)\tfrac{1}{3} + (a' - a)(b' - b)\tfrac{1}{4}]\frac{(e' - e)}{R}\Big]$$

$$= \frac{l}{6}\Big[ab + \tfrac{1}{2}ab' + \tfrac{1}{2}a'b + a'b'$$

$$+ (ab + \tfrac{1}{2}ab' + \tfrac{1}{2}a'b + a'b')\frac{e}{R}$$

$$+ (\tfrac{1}{4}ab + \tfrac{1}{4}ab' + \tfrac{1}{4}a'b + \tfrac{3}{4}a'b')\frac{e' - e}{R}\Big]$$

$$=\frac{l}{6}\left[(A+4A_m+A')+(A+4A_m+A')\frac{e}{R}+(2A_m+A')\frac{e'-e}{R}\right]$$

Volume

$$=\frac{l}{6}(A+4A_m+A')+\frac{l}{6R}\left[(A+2A_m)e+(2A_m+A')e'\right] \quad (5)$$

where A_m is the area of the mid section, and e, e', are positive for the convex side of the curve.

The first product is the volume which would be obtained by the prismoidal formula neglecting curvature, leaving the second for the correction due to curvature.

Volume for Railway Earthwork

IN CUBIC YARDS BY AVERAGING END AREAS.

---

6. Formula for Three-level Ground. —Dimensions in feet, volume in cubic yards.

$$\text{Area of section} = (a+c)\frac{d_1}{2} + (a+c)\frac{d_r}{2} - \frac{a\,b}{2}$$

$$= (a+c)\frac{w}{2} - \frac{a\,b}{2}$$

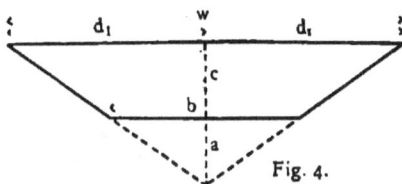

Fig. 4.

At the next section, 100 feet from the first, priming corresponding parts,

$$\text{Area}' = (a+c')\frac{w'}{2} - \frac{a\,b}{2}$$

Multiplying the average area by the length, and dividing by 27 to reduce to cubic yards,

$$\text{Volume} = (\text{area} + \text{area}')\frac{100}{2 \times 27}$$

$$\text{Volume} = (a+c)\,w\tfrac{50}{54} - ab\tfrac{50}{54} + (a+c')\,w'\tfrac{50}{54} - ab\tfrac{50}{54}. \qquad (6)$$

I. e., the effect of each cross section on the volume reduces to *the constant times the total width between slope stakes times the sum formed by adding the center height to the height of the grade triangle; minus the constant times the base times the altitude of the grade triangle.*

For the next solid,

$$\text{Volume}' = (\text{area}' + \text{area}'')\tfrac{100}{54}.$$

The terms comprising *area'* should therefore be added separately, for convenience in finding the second volume, etc.

7. **Five-level Ground.** —In five-level ground the intermediate heights, t_1 and t_r, are taken over (or under) the edges of the roadbed. By joining the edges of the roadbed with the ground at the center, two triangles are formed on the left having t_1 for the common base, d_1 for

7

the sum of the altitudes, and $t_1 d_1$ for double area ; two on the right with double area $t_r d_r$; and one at the center with double area bc.

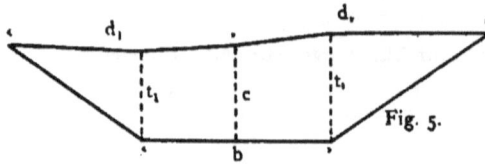

Fig. 5.

Therefore,   Area first end $= \dfrac{t_1 d_1}{2} + \dfrac{t_r d_r}{2} + \dfrac{bc}{2}$

Area second end $= \dfrac{t_1' d_1'}{2} + \dfrac{t_r' d_r'}{2} + \dfrac{bc'}{2}$

Volume $= t_1 d_1 \frac{l_0}{5 4} + t_r d_r \frac{l_0}{5 4} + c b \frac{l_0}{5 4} + t_1' d_1' \frac{l_0}{5 4} + t_r' d_r' \frac{l_0}{5 4} + c' b \frac{l_0}{5 4}.$ (7)

I. e., the effect of each cross section on the volume reduces to *the constant times the distance out to the left slope stake times the left shoulder height ; plus the constant times the distance out to the right slope stake times the right shoulder height; plus the constant times the width of roadbed times the center height.*

8. Irregular Ground.—In Fig. 6, let $d_1, f_1, g_1 \ldots,$ and d_r, f_r, g_r $\ldots =$ the distances to the left, and right, to the slope stakes, and intermediate breaks in the ground surface; and $r_1, u_1, t_1 \ldots,$ and $r_r, u_r,$ $t_r \ldots, =$ the corresponding heights.

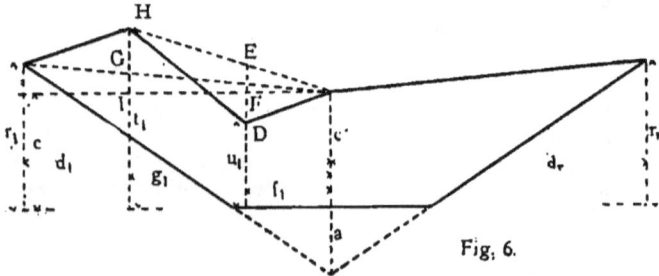

Fig. 6.

The area of the section may be found, by taking the two triangles with the common base $a + c$, and altitudes d_1 and d_r, adding the triangle with area $GH \cdot \dfrac{d_1}{2}$, and subtracting the triangle with area $DE \cdot \dfrac{g_1}{2}$, and the grade triangle $\dfrac{ab}{2}$. This gives,

Area first end (the right side being regular)

$= (a + c) \dfrac{d_1}{2} + \overline{GH} \cdot \dfrac{d_1}{2} - \overline{DE} \cdot \dfrac{g_1}{2} \pm \text{etc.} + (a + c) \dfrac{d_r}{2} - \dfrac{ab}{2}$

$= (a + c) \dfrac{d_1}{2} + (t_1 - c - \overline{IG}) \dfrac{d_1}{2} + (u_1 - c - \overline{FE})_\frac{g_1}{2} + \text{etc.} + (a + c) \dfrac{d_r}{2} - \dfrac{ab}{2},$

Eq. 8]   *IRREGULAR GROUND.*   9

where IG and FE are to be taken as negative if measured below the horizontal line FI. Expressing them in terms of measured quantities,

since $IG = (r_1-c)\dfrac{g_1}{d_1}$ and $FE = (t_1-c)\dfrac{f_1}{g_1}$,

Area first end

$$= (a+c)\frac{d_1}{2} + \left[t_1-c-(r_1-c)\frac{g_1}{d_1}\right]\frac{d_1}{2} + \left[u_1-c-(t_1-c)\frac{f_1}{g_1}\right]\frac{g_1}{2} + \text{etc.}$$

$$+ (a+c)\frac{d_r}{2} - \frac{ab}{2}$$

$$= (a+t_1)\frac{d_1}{2} + (u_1-r_1)\frac{g_1}{2} + (c-t_1)\frac{f_1}{2} + \text{etc.} \; + (a+c)\frac{d_r}{2} - \frac{ab}{2},$$

where each break or intermediate height H, D, etc., adds a term of the form $(u_1-r_1)\dfrac{g_1}{2}$. If H is omitted, the corresponding term $(u_1-r_1)\dfrac{g_1}{2}$ disappears and the other term becomes $(c-r_1)\dfrac{f_1}{2}$, as may be found by leaving off the triangle $GH\cdot\dfrac{d_1}{2}$; if H and D are omitted both terms disappear, and $(a+t_1)\dfrac{d_1}{2}$ becomes $(a+c)\dfrac{d_1}{2}$, as may be seen from the expression for the right side.

If the next cross section, 100 feet distant, is similar to the first, and the same letters with primes are used for corresponding parts, we shall have,

Area second end,

$$= (a+t_1')\frac{d_1'}{2} + (u_1'-r_1')\frac{g_1'}{2} + (c'-t_1')\frac{f_1'}{2} + \text{etc.} \; + (a+c')\frac{d_r'}{2} - \frac{ab}{2}.$$

The volume will be found by multiplying the average area by the length, and dividing by 27 to reduce to cubic yards.

$$\text{Volume in cubic yards} = (\text{area} + \text{area}')\,\frac{100}{2 \times 27}.$$

Substituting the value of each end area,

Volume

$$= (a+t_1)d_1\tfrac{50}{54} + (u_1-r_1)g_1\tfrac{50}{54} + (c-t_1)f_1\tfrac{50}{54} + \text{etc.} \; + (a+c)d_r\tfrac{50}{54} - ab\tfrac{50}{54}$$
$$+ (a+t_1')d_1'\tfrac{50}{54} + (u_1'-r_1')g_1'\tfrac{50}{54} + (c'-t_1')f_1'\tfrac{50}{54} + \text{etc.} \; + (a+c')d_r'\tfrac{50}{54} - ab\tfrac{50}{54}$$

$$(8)$$

Or, the effect of each cross section on the volume reduces:

For the irregular side; to a constant times the distance to the slope stake times the sum formed by adding the height next the slope stake to that of the grade triangle; plus the constant times the distance out to each break, in turn, times the difference formed by subtracting the first height outside from the first height inside that break.

For the regular side; to the constant times the distance out to the slope stake times the sum formed by adding the center height to the height of the grade triangle.

From the total sum must be subtracted the constant times the base times the altitude of the grade triangle.

For the next solid,

$$\text{Volume} = (\text{area}' + \text{area}'')\, \frac{100}{2 \times 27} ;$$

the terms comprising *area'* should therefore be added separately, for convenience in finding the second volume, etc.

It may be noted that, if the expressions for the areas of the original triangles had not been reduced, the constant times the base times the altitude of each, would appear as its effect upon the volume. Hence *the computer may divide the sections into any triangles he may wish;* the volume will result by multiplying the constant into the base times the altitude of each triangle, and adding the products.

9. **Tables.**—The table gives the value of $(a+c)\,d\,\frac{55}{88}$ (or $t\,d\,\frac{55}{88}$), for $a+c$ (height) and d (width) as arguments; or it gives the effect which any triangle in the cross section has upon the volume, for a length between stations of 100 feet, by using the altitude and base of the triangle as arguments. After looking out the partial volume for each triangle for both ends and adding, if the length be not 100, multiply the sum by $\frac{1}{100}$ of the length.

To look out the partial volume for a triangle:—Find the integral part of the height factor in the line of height headings at the top of the page; then move down this column until opposite the integral part of the width factor in the width column and note the number; then move to the left on this line (or the next below if the tenths of width be greater than 5) and take out the " Correction for tenths of height "; them move to the bottom of the page and under the column first used (or the next to the right if the tenths of height be greater than 5) take out the "Correction for tenths of width." Take the sum of the three mentally. Look out the quantity for each of the other terms; the algebraic sum will be the total volume. The terms coming from "area second end," should be kept separate as they will be used in finding the volume of the next solid.

If one of the arguments cannot be found in the table divide it by two, look out the number and multiply by two; if both cannot, divide each by two and multiply by four.

In order to impress upon the mind the fact that the correction for tenths of width is placed under the height column, it should be remembered that the differences down the column are constant, so that corrections or proportional parts of these differences can be placed under it, but the widths are increased as we move down the column, or these corrections are for tenths of width. Similarly for the lines horizontally, or the corrections for tenths of height.

After becoming familiar with the table, it will often save time to call height "width," and width "height," and to take the correction for tenths from the side or bottom of the page as most convenient, *provided* always that it is opposite (or under) the heading number to which it belongs.

Eq. 8]   *EXAMPLE, FIVE-LEVEL GROUND.*   11

10. **Example, Three-level Ground.—**

Sta.	Center.	Left.	Right.
20 . . . . . 10.8, ...		$\dfrac{8.4^*}{19.6}$, ...	$\dfrac{11.4}{24.1}$
21 . . . . . 24.2, ...		$\dfrac{11.1}{23.7}$, ...	$\dfrac{32.6}{55.9}$
21 + 20 . . 46.7, ...		$\dfrac{20.0}{37.0}$, ...	$\dfrac{50.6}{82.9}$

$a = 4.7, \, b = 14.$ †

Substituting in formula (6), or following the rule § 6, we shall have, in connection with § 9 :

Sta. 20,

$a + c = 4.7 + 10.8 = 15.5, \; w = 19.6 + 24.1 = 43.7.$

From the Table of Triangular Prismoids,

Under height 15, right of width 43, ...	597
Left of width 44, correction for 0.5, ...	20
Under height 15, correction for 0.7, ...	10
	627
Arguments, $a = 4.7$ and $b = 14$ (grade prism), ...	—61 566

Sta. 21,

$a + c' = 4.7 + 24.2 = 28.9, \; w' = 23.7 + 55.9 = 79.6, \; . . .$	2131
$a = 4.7, \; b = 14, \; . . . .$	—61 2070
Total in yards, ...	2636
From above, for sta. 21, ...	2070

Sta. 21 + 20,

$a + c' = 4.7 + 46.7 = 51.4, \; w' = 37.0 + 82.9 = 119.9, \; 5712$	
$a = 4.7, \; b = 14$ (grade prism), ...	—61 5651
	7721
Total in yards . . . . 7721 × .20 =	1544
Grand total, ...	4180

11. **Example, Five-level Ground.—**

Sta.	Center.	Left.		Right.	
46 . . . . 10.8, . . .		$\dfrac{11.4}{7}$.	$\dfrac{12.1}{25.1}$, ...	$\dfrac{9.9}{7}$ . .	$\dfrac{8.6}{19.9}$
47 . . . . . 6.4, . . .		$\dfrac{6.8}{7}$.	$\dfrac{7.2}{17.8}$, ...	$\dfrac{5.7}{7}$ . .	$\dfrac{5.0}{14.5}$

---

* The numerator is the height above or below grade at the point and the denominator is its distance from the center.

† The side slope s is here assumed to be one and one-half horizontal to one vertical, so that

$$a = \frac{b}{2} \div s = \frac{14}{2} \div \frac{1\frac{1}{2}}{1} = 4.7.$$

Substituting in formula (7), using the Table for the multiplications,

Sta. 46, arguments, 25.1 and 11.4, 265
"   19.9 "   9.9, 183
"   14.0 " 10.8, 140 588
                                    ——

Similarly for Sta. 47,   272
                                            ——

Total in yards,   860

12. Example, Irregular Ground.—

| Sta. | Center. | Left. | | | | Right. | |

20 . . . 10.8, ... $\dfrac{8.4}{19.6}$ . . . . . $\dfrac{11.2}{18}$ . . $\dfrac{11.4}{24.1}$

21 . . . 24.2, . . $\dfrac{18.6}{16}$. $\dfrac{11.1}{23.7}$, ... $\dfrac{20.4}{42}$ $\dfrac{32.6}{55.9}$

21+20 . 46.7, . . $\dfrac{40.7}{20}$ $\dfrac{20.0}{37.0}$, . . . $\dfrac{48.1}{20}$ . . $\dfrac{50}{58}$. $\dfrac{50.6}{82.9}$

$b = 14$, $s = 1.5$ to 1, $a = 4.7$.

Substituting in formula (8), using the Table for the multiplications,
Volume from sta. 20 to sta. 21.

Sta. 20,

$a+l_r=4.7+11.2=15.9$, $d_r=24.1$; $c-r_r=10.8-11.4=-0.6$, $g_r=18$;
$a+c=15.5$, $d_l=19.6$; $a=4.7$, $b=14$.

Under height 15, right of width 24, ... 333
Left of width 24, cor. for 0.9, ... 20
Under height 16, cor. for 0.1, ... 1
                                            ——
                                            354
Arguments, —0.6 and 18, ... —10
"   19.6 and 15.5, ... 281
"   14 and 4.7 (grade prism), ... —61
                                            ——   564

Sta. 21, Arguments,
"   20.4+4.7=25.1, and 55.9, ... 1299
"   24.2—32.6=—8.4, and 42, ... —327
"   18.6+4.7=23.3, and 23.7, ... 512
"   24.2—11.1=13.1, and 16, ... 194
"   14 and 4.7 (grade prism), ... —61
                                            ——   1617

Total in yards, ...   2181

Volume from 21 to 21+20.

From above for sta. 21, . .   1617

Sta. 21+20,
            50+4.7=54.7, and 82.9, ... 4200
            48.1—50.6=—2.5, and 58, ... —134
            46.7—50=—3.3, and 20, ... —62
            40.7+4.7=45.4, and 37, ... 1556
            46.7—20=26.7, and 20, ... 494
            14 and 4.7 (grade prism), ... • —61
                                            ——   5993

                                            7610

7610 × .20 =   1522
                                            ——
                                            3703

Eq. 8]   SIDEHILL WORK.   13

Any other method of dividing the sections into triangles can be used if preferred, although it will result in increased labor.

13. **Borrow Pits.**—In borrow pits, ditches or other work, where the cross sections are made up of triangles or trapezoids, take the altitude and base of each triangle, or the sum of the parallel sides, and the perpendicular distance between them for each trapezoid, as arguments, as with the two sections below.

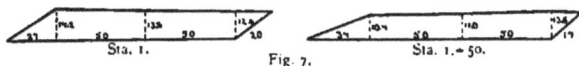

Fig. 7.

$$
\begin{array}{llll}
\text{Sta. 1, 27 and 14 2, .} & & 355 \\
\quad\quad 50 \;\; `` \;\; 28.0, . \; . & 1296 \\
\quad\quad 50 \;\; `` \;\; 26.4, . & 1223 \\
\quad\quad 20 \;\; `` \;\; 12.6, . \; . & 233 \\
& & \overline{\quad\quad} \\
& & 3107 \\
\text{Similarly for sta. 1+50,} & 2672 \\
& & \overline{\quad\quad} \\
& & 5779
\end{array}
$$

5779 × .50 = 2890 yards.

14. **Sidehill Work.**—When a "grade point" occurs in a cross section, one side being in cut, the other in fill, to compute the cutting, put in a break at the grade point and call the cuts on the fill side zero, and the distance to the slope stake $b \div 2$. Then proceed as for irregular sections.

In computing fills the process would be reversed.

EXAMPLE.

Sta.	Center.	Left.		Right.
46 . . −2.4, .		$\dfrac{0.0}{4.3}$	$\dfrac{2.2}{12.3}$, . .	. . . . $\dfrac{-6.4}{16.6}$
47 . . +1.7, ...		$\dfrac{3.8}{14.7}$	$\dfrac{0.0}{2.6}$ . .	$\dfrac{-2.9}{11.4}$

Roadbed 18 feet in cut, 14 feet in fill, slopes 1½ to 1. Altitude of grade triangle in cut $= 9 \div \frac{1\frac{1}{2}}{1} = 6$, in fill $= 7 \div \frac{1\frac{1}{2}}{1} = 4.7$.

For cutting, the notes would be

46 . : . 0.0, . . .	$\dfrac{0.0}{4.3}$ . . .	$\dfrac{2.2}{12.3}$, ...	$\dfrac{0.0}{9.0}$
47 . . . 1.7, ...	. . $\dfrac{3.8}{14.7}$, . . .	$\dfrac{0.0}{2.6}$	$\dfrac{0.0}{9.0}$

Giving,

$$
\begin{array}{ll}
\text{Sta. 46, 0.0+6.0= 6, and 12.3, . .} & 69 \\
\quad\quad 0.0-2.2=-2.2, \text{ and } 4.3, . & -9 \\
\quad\quad 0.0+6.0= 6, \text{ and } 9, . \; . \; . & 50 \\
\quad\quad \text{Grade triangle, 6 and 18, } & -100 \\
& \overline{\quad\quad} \\
& - 10 \\
\text{Sta. 47, similarly, . . } . \; . \; . \; . \; . & 60 \\
& \overline{\quad} \\
\quad\quad\quad \text{Total in cu. yards, . . .} & 70
\end{array}
$$

Or it may be more convenient to leave out the grade prism, thus :

Sta. 46, height$=$2.2, width$=$ 9 $-$4.3$=$ 4.7, giving . .   5   9
"   47,   " $=$1.7,   " $=$14.7$+$2.6$=$17.3,   "   . . 28
        " $=$3.8,   "   . . . . 9.0.   "   . . 32
                                                — 60

Total in yards, ...   69

For filling, the notes would be

Sta.	Center.	Left.		Right.
46 . .	—2.4, . . .	$\frac{0.0}{4.3}$ . . .	$\frac{0.0}{7.0}$ . . .	. . . $\frac{-6.4}{16.6}$
47 . . .	0.0, ...	$\frac{0.0}{7.0}$ . . .	$\frac{0.0}{2.6}$ . . .	$\frac{-2.9}{11.4}$

Giving,

Sta. 46,   0.0$+$4.7$=$4.7, and   7, . .   31
            2.4$-$0.0$=$2.4, "   4.3, . .   9
            2.4$+$4.7$=$7.1, " 16.6, . . 109
        Grade triangle, 4.7 " 14, . . $-$61
                                        —— 88
Sta. 47, similarly, ...   11

Total fill in cu. yards, . . .   99

15. **End Sections.**—A cross section should be taken wherever either edge of the roadbed in cutting reaches grade. This will require two sections in passing from a cut to a fill, or *vice versa*, while a third is sometimes added where the center line reaches grade. The solid, in cut, between the two sections, is a pyramid with volume equal to area of base into one-third the altitude, or it may be taken under the rule, base into one-half the altitude, and afterwards corrected by the prismoidal formula.

The solid in fill, between the two sections, can usually be taken as having the same altitude, the volume not being required as accurately as for cut; the exact values can be interpolated when necessary.

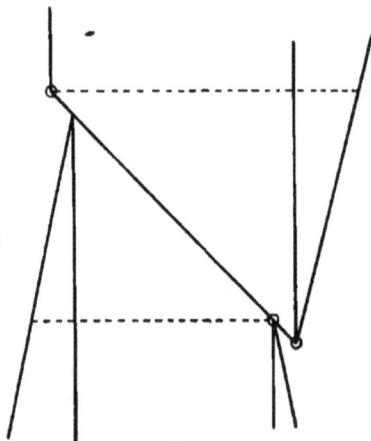

Fig 8.

16. **Correction for Curvature.**—For a solid corresponding to the triangular prismoid, as described in § 5, we have from (5),

Eq. 13] *CORRECTION FOR CURVATURE.* 15

$$\text{Correction} = \frac{l}{6R}\left[(A + 2A_m)e + (2A_m + A')e'\right] \qquad (9)$$

$$= \frac{l}{2R}\left[A e + A' e'\right] \text{ (approx.)} \qquad (10)$$

The latter equation is the one in common use, giving, with less labor, sufficient accuracy for earthwork solids.

Three-level Ground.—To find e, the eccentricity of the center of gravity, we have,

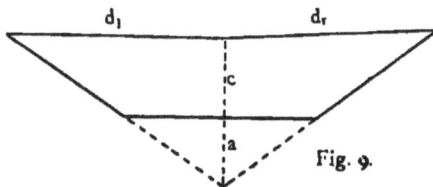

Fig. 9.

For the right side,

$$\text{Area} = (a + c)\frac{d_r}{2}$$

$$\text{Center of gravity from center line} = \frac{d_r}{3}$$

$$(\text{Product})_r = (a + c)\frac{d_r^2}{6}$$

Similarly, for the left side,

$$(\text{Product})_l = -(a + c)\frac{d_l^2}{6}$$

Hence

$$e = \frac{(\text{product})_r + (\text{product})_l}{\text{total area}} = \frac{(a+c)\dfrac{d_r^2 - d_l^2}{6}}{(a+c)\dfrac{d_r + d_l}{2}},$$

Or,

$$e = \frac{d_r - d_l}{3} \qquad (11)$$

$$\text{Correction} = \frac{l}{6R}\left[A(d_r - d_l) + A'(d_r' - d_l')\right] \qquad (12)$$

where A and A' must include the grade triangle.

Or,

$$\text{Correction}_y = \frac{1}{3R}\left[Q(d_r - d_l) + Q'(d_r' - d_l')\right] \qquad (13)$$

where Q and Q' are the tabular quantities or volumes corresponding to A and A' ($A = \frac{27}{50}Q$) and Correction $_y$ is in cubic yards.

Five-level and irregular ground can usually be taken as three-level, with the same distances to slope stakes, for the purposes of this correction, unless the intermediate heights show a marked tendency to lie above, or below, the lines joining the side heights with the center, when the section can be plotted and an equalizing line drawn, or in extreme cases each triangular prismoid can be taken separately by (10).

Sidehill Ground, neglecting all breaks between slope stakes. The center of gravity will be on the line joining the center of the base with the vertex, at one-third its length from the base.

The distance from the lower end of this line to the center, Fig. 10,

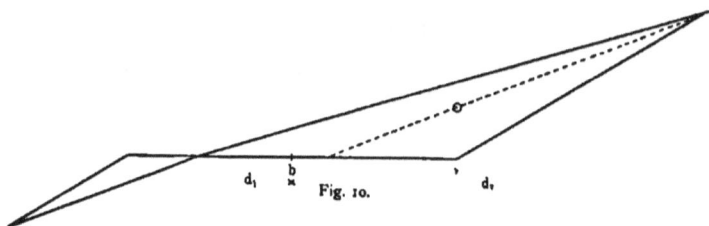

Fig. 10.

$$=\frac{b}{2}-\left(\frac{b}{2}+d_1\right)\tfrac{1}{2}=\frac{b}{4}-\frac{d_1}{2}$$

Distance from the upper end $= d_r$.

Hence $\quad e=\tfrac{2}{3}\left(\frac{b}{4}-\frac{d_1}{2}\right)+\tfrac{1}{3}d_r=\tfrac{1}{3}\left(\frac{b}{2}-d_1+d_r\right)$ (14)

where d_1 will be positive if the grade point is on the same side of the center as the slope stake.

Substituting in (10),

$$\text{Correction} = \frac{l}{6R}\left[A(\frac{b}{2}-d_1+d_r)+A'(\frac{b}{2}-d_1'+d_r')\right]$$ (15)

$$\text{Correction}_y = \frac{1}{3R}\left[Q(\frac{b}{2}-d_1+d_r)+Q'(\frac{b}{2}-d_1'+d_r')\right]$$ (16)

17. **Example, Correction for Curvature.**—In the example of § 12, let the center line be on a 7° curve to the right, $R = 5730 \div 7 = 819$, and let the intermediate breaks be omitted in finding the correction for curvature.

Volume from sta. 2) to sta. 21.

Q, including the grade triangle, $= 625$; $Q' = 1678$; $d_1 = 19.6$; $d_r = 24 1$; $d_1' = 23.7$; $d_r' = 55.9$.

Substituting in (13),

$$\text{Correction} = \frac{1}{3\times819}[625(24.1-19 6)+1678(55.9-23.7)] = \quad 23$$

Volume from sta. 21 to sta. 21 + 20,

$$\text{Correction} = \frac{1}{3\times819}[1678(55.9-23.7)+6054(82.9-37.0)] = 135$$

$$.2\times135 \qquad\qquad = \quad 27$$

Total in cu. yards $= \quad -50$

The correction is negative, the center of gravity and the center of the curve lying on the same side of the center line.

In the example of § 14, let the center line be on a 7° curve to the right. Volume from sta. 46 to sta. 47.

Substituting in (16),

$$\text{Correction} = \frac{1}{3\times819}[10(9-12.3-4.3)+60(9-14.7+2.6)] = 0$$

This correction would be positive, the center of gravity and the center of the curve being on opposite sides of the center line.

Prismoidal Correction in Cubic Yards.

18. **Three-level Ground.**—The prismoidal correction to be applied to the volume of a triangular prismoid found by averaging end areas in order to obtain the true volume, has been found to be by (2),

$$\text{Correction} = (c - c')(d' - d)\frac{l}{12}$$

where c and d are the altitude and base of the triangle at the first end, and c' and d' those at the second end.

For three-level ground, per station in cubic yards, this would reduce to

$$\text{Pris. correction} = (c - c')(w' - w)\frac{50}{3 \times 54} \qquad (17)$$

where, as in Fig. 4 and eq. (6), c and c' are the center heights and w and w' are the total widths.

This correction is to be added to the approximate volume found by end areas; it disappears when the center heights, or the widths, are equal, and it can often be neglected without computation when inspection shows one of the factors very small.

19. **Five-level Ground.**—The intermediate heights having been taken over (or under) the edges of the roadbed. In § 7, see Fig. 5, each section was divided into three triangles, one having the distance to the left slope stake, d_1, for base, and the height over the left edge of the roadbed, t_1, for altitude, the second the corresponding dimensions, d_r, t_r, on the right, and the third the width of roadbed and the center height. Applying (2), and remembering that since the width of roadbed is the same at both ends, the correction for the third triangle will be zero:

$$\text{Pris. correction} = \left[(t_1 - t_1')(d_1' - d_1) + (t_r - t_r')(d_r' - d_r) \right]\frac{50}{3 \times 54} \qquad (18)$$

20. **Irregular Ground, Borrow Pits, etc.**—Since the areas of the sections will be made up of triangles, the correction will be made up of as many terms as there are triangular prismoids in the volume, each term being of the form

$$(c - c')(w' - w)\frac{50}{3 \times 54}.$$

This requires that there should be the same number of breaks at each end of the solid, as already indicated in § 1, in deriving the prismoidal formula and defining the prismoid. If there are not, the distance from the center at which a ridge or hollow runs out at the more regular section should be noted in the field, and the height either noted or computed.

17

A. M. Wellington* recommends that for the *purposes of the prismoidal correction* irregular ground be assumed to be three-level ground having the center height and total width of the irregular section.

This makes the correction the same as for regular ground, and while for a single volume it may give a result considerably in error, the sign is as likely to be plus as minus, so that there is less tendency to accumulation of error than in the various approximate methods described in §§ 2-4

21. Table of Prismoidal Corrections.—The table gives the value of $(c - c')(w' - w)\dfrac{50}{3 \times 54}$, for $c-c'$ (height — height') and $w' - w$ (width' —width) as arguments ; which is the correction for a length of 100 feet. If the length be not 100, multiply the result by $\frac{1}{100}$ of the length.

To look out the number :—Find the integral part of the height factor in the line of height headings at the top of the page ; then move down this column until opposite the integral part of the width factor in the width column, and note the number ; then move to the left on this line (or the next below if the tenths of the width factor be greater than 5) and take out the "Correction for tenths of height—height' " ; then move to the bottom of the page and under the column first used (or the next to the right if the tenths of the height factor be greater than 5) and take out the "Correction for tenths of width'—width." Take the sum of the three mentally.

If one of the arguments cannot be found in the table, divide it by two, look out the number and multiply by two ; if both cannot, divide each by two and multiply by four.

The remarks of § 9 apply to this table the same as to the table of triangular prismoids.

22. Example, Three-level Ground.—See § 10.

Sta.	Center.	Left.	Right.
20 . . . . .10,8, ...		$\dfrac{8.4}{19.6}$, ...	$\dfrac{11.4}{24.1}$
21 . . . . .24.2, ...		$\dfrac{11.1}{23.7}$, ...	$\dfrac{32.6}{55.9}$
21+20 . . .46.7, ...		$\dfrac{20.0}{37.0}$, ...	$\dfrac{50.6}{82.9}$

Correction for volume from sta. 20 to sta. 21.

$c - c' = 10.8 - 24.2 = -13.4$, $w' - w = 79.6 - 43.7 = +35.9$.

From the Table of Prismoidal Corrections.

Under 13, right of argument 35,		= 140	
Left	"	36, corr. for 0.4 =	4
Under		13, corr. for 0.9 =	4	— 148

* *Computation from Diagrams of Railway Earthwork.*

Eq. 18]   *EXAMPLE, IRREGULAR GROUND.*   19

Correction for volume from sta. 21 to sta. 21 + 20.

$c - c' = 24.2 - 46.7 = -22.5$, $w' - w = 119.9, -79.6 = +40.3$;

22.5 and 40.3 as arguments $= 272 + 6 + 2 = 280$

$$-280 \times .20 = \qquad\qquad\qquad\qquad -56$$

Total correction to the volume found by " End areas " $= -204$

23. **Example, Five-level Ground.**—

Sta.	Center.	Left.		Right.	
46	. . . 10.8,	. . $\frac{11.4}{8}$ . .	$\frac{12.1}{20.1}$,	. . . . $\frac{9.9}{8}$	. $\frac{8.6}{16.6}$
47	. . . . 6.4,	. . $\frac{6.8}{8}$ . .	$\frac{7.2}{15.2}$,	. . . . $\frac{5.7}{8}$	. $\frac{5.0}{13.0}$

$b = 16$, side slope ratio $s = 1$ to 1. See Fig. 5 and (18).

$l_1 - l_1' = 11.4 - 6.8 = +4.6$, $d_1' - d_1 = 15.2 - 20.1 = -4.9$.

$l_r - l_r' = 9.9 - 5.7 = +4.2$, $d_r' - d_r = 13.0 - 16.6 = -3.6$.

4.9 and 4.6 as arguments $= 5 + 1 + 1 = -7$

4.2 and 3.6   " $= 4 + 1$ $= -5$

Correction in yards $= -12$

By the approximate method of § 20: $10.8 - 6.4 = +4.4$; $28.2 - 36.7 = -8.5$.

$+4.4$ and $-8.5 = 10 + 1 + 1 = -12$, $=$ correction in yards.

24. **Example, Irregular Ground.**—See § 12.

Sta.	Center.	Left.			Right.	
20	. . . 10.8,	. . . . . .	$\frac{8.4}{19.6}$,	. . . . . .	$\frac{11.2}{18}$	. . . $\frac{11.4}{24.1}$
21	. . 24.2,	. . . $\frac{18.6}{16}$ . .	$\frac{11.1}{23.7}$,	. . . . . .	$\frac{20.4}{42}$	. . . $\frac{32.6}{55.9}$
21+20	. 46.7,	. . . $\frac{40.7}{20}$ . .	$\frac{20.0}{37.0}$, . . .	$\frac{48.1}{20}$ . .	$\frac{50.0}{58}$	. . . $\frac{50.6}{82.9}$

$b = 14$, $s = 1.5$ to 1, $a = 4.7$.

The break 16 feet out at left of sta. 21 runs out 10 feet left, at sta. 20. The break 20 feet out at right of sta. 21 + 20 runs out at the center at sta. 21.

By interpolation, the height at 10 feet to the left at sta. 20 = 10.8 —

$$\frac{10}{19.6}[10.8 - 8.4] = 9.6.$$

Correction for volume from sta. 20 to sta. 21.

Sta. 20 would be written

10.8,	. . . $\frac{9.6}{10}$	. . . $\frac{8.4}{19.6}$,	. . . . . .	$\frac{11.2}{18}$	. . . $\frac{11.4}{24.1}$

Changing the factors of § 12 which were used in finding volume, or the areas of the triangles, to correspond with the interpolated break,

Arguments, 19.6−23.7=− 4.1, 23.3−14.3=+ 9.0, corr.=− 11
"   24.1−55.9=−31.8, 25.1−15.9=+ 9.2, "   − 90.
"   10 −16 =− 6.0, 13.1− 2.4=+10.7, "   − 20
"   18 −42 =−24.0, −8.4+0.6=− 7.8, "   + 58 −63

Correction for volume from sta. 21 to sta. 21+20.
Sta. 21 would be written

$$24.2, \ldots \frac{18.6}{16} \ldots \frac{11.1}{23.7}, \ldots \frac{24.2}{0} \ldots \frac{20.4}{42} \ldots \frac{32.6}{55.9},$$

giving as above,

Arguments, 23.7−37.0=−13.3, 45.4−23.3=+22.1, corr.=− 90
"   55.9−82.9=−27.0, 54.7−25.1=+29.6, "   −247
"   16 −20 =− 4.0, 26.7−13.1=+13.6, "   −.17
"   42 −58 =−16.0,−2.5+8.4=+ 5.9, "   − 29
"   0 −20 =−20.0,−3.3− 3.8=− 7.1, "   + 44

                                            −339
339×.20=   −68

           Total correction in yards=   −131

By the approximate method of § 20 the result would be that found for the first example, or − 204 cu. yds.

25. **Example, Borrow Pits.**—See example, § 13.

Arguments which were used in finding volume; for sta. 1, 27 and 14.2, 50 and 28.0, 50 and 26.4, 20 and 12.6; for sta. 1 + 50, 34 and 10.4, 50 and 21.4, 50 and 24.6, 17 and 13.6.

The corrections will all disappear except for the first and last product at each station.

Arguments, 27−34=− 7, 10.4−14.2 =− 3.8, corr. =+ 8
"   20−17=+ 3, 13.6−12.6=+1.0, "   =+ 1

                                         9

Correction in yards = 9 + .50 = 5.

Computation---End Areas with Prismoidal Correction.

26. Three-level Ground.—

Sta. Center.		Left.	Right.
0 . . 7.1	. . . .	$\frac{6.4}{14.4}$	$\frac{8.0}{16.0}$
1 . . 9.8	. . . .	$\frac{7.6}{15.6}$	$\frac{12.4}{20.4}$
2 . . 12.4	. . . .	$\frac{12.4}{20.4}$	$\frac{10.3}{18.3}$
3 . . 8.7	. . . .	$\frac{9.8}{17.8}$	$\frac{8.1}{16.1}$
4 . . 10.2	. . . .	$\frac{11.6}{19.6}$	$\frac{9.8}{17.8}$

Roadbed 16 ft., side slopes 1 to 1.

COMPUTATION.

Sta.	Width.	Height, or h.+a.	Yds. by End areas		w — w'.	ht' — ht.	Yards. pris'dl.
0	30.4	15.1	426				
1	36.0	17.8	594	782	—5.6	+2.7	—5
2	38.7	20.4	731	1087	—2.7	+2.6	—2
3	33.9	16.7	525	1018	+4.8	—3.7	—6
4	37.4	18.2	631	918	—3.5	+1.5	—2

Approx. volume = 3805   — 15
Prismo'l cor. = — 15

True volume = 3790

27. Irregular Ground.—

9 . . 6.8	. . . .	$\frac{5.2}{13.2}$.	$\frac{7.4}{8}$.	$\frac{3.4}{11.4}$	
10 . . 4.9	. . . .	$\frac{9.6}{17.6}$.	$\frac{4.1}{9}$.	$\frac{3.6}{11.6}$	
11 . . 9.0 . . $\frac{7.2}{8}$ . .	$\frac{5.0}{12}$ . .	$\frac{8.4}{16.4}$.	$\frac{5.2}{7}$.	$\frac{8.9}{16.9}$	$\frac{10,14}{16}\ \Big\vert\ \frac{*}{12}$
12 . . 6.8	$\frac{11.4}{16}$ . .	$\frac{12.8}{20.8}$	. . . .	$\frac{7.1}{15.1}$	
13 . . 8.6	$\frac{7.7}{12}$ . .	$\frac{7.4}{15.4}$	. . . .	$\frac{5.2}{13.2}$	

*The breaks at the left of 11 run out 10 ft. and 14 ft. respectively to the left of 10, etc.

$+47$ . . 2.0 . . . . $\dfrac{4}{10}$ . . $\dfrac{5.4}{13.4}$ . . . . . $\dfrac{0.0}{8.0}$

14 . —5.1 . . . . . . $\dfrac{0.0}{4.0}$ . . $\dfrac{4.2}{12.2}$ . . . . . $-\dfrac{2.0}{10.0}$ . . . $\dfrac{|}{8\,|}$

$+20$. —6.0 . . . . $\dfrac{0.0}{8.0}$ . . . . $-\dfrac{4.4}{13.6}$

Roadbed 16 ft., slopes 1 to 1 in cut.
Roadbed 14 ft., slopes 1½ to 1 in fill.

Computation on the assumption that, *for the purpose of the prismoidal correction only,* the ground is regular.

Sta.	Width.	Height, or h.+a.	Yds. by End areas	w—w'.	ht'—ht.	Yards. pris'dl.
9	13.2	14 8	182			
	11.4	15.4	163			
	8	3.4	25			
10	17.6	12.9	211			
	11.6	12.1	130			
	9	1.3	11	484	—4.6	—1.9	+2
11	16.4	13.0	198			
	12.	—1.2	—13			
	8	+4	30			
	16.9	13.2	207			
	7	+0.1	1	537	—4.1	+4.1	—5
12	20.8	19.4	374			
	16.0	—6.0	—89			
	15.1	14.8	206	676	—2.6	—2.2	+1
13	15.4	15.7	224			
	12.	1.2	13			
	13.2	16.6	203	693	+7.3	+1.8	+4
+47	13.4	12.0	148			
	10.	—3.4	—32			
	8	10.0	74	184	+7.2	—6.6	—7
Pris'l for 14 {	16	5.4				
14	4.0	4.2	16	46	+12.0	—1.2	—3
+20	0.0	0.0	00	3	+4.0	—4.2	—1

Approx. volume = 2623   —16
Prismo'l cor. = —   9   + 7
Corrected volume =   2614   — 9

Computation—*With the true prismoidal correction.*

Sta.	Width.	Height, or h.+a.	Yds. by End areas		w−w'.	ht'−ht.	Yards, pris'dl.
9	13.2	14.8	182				
	11.4	15.4	163				
	8	3.4	25				
10	17.6	12.9	211		−4.4	−1.9	+2
Pris'l* for 11	17.6	16.6					
	14	−1.9					
	10	−3.7					
	11.6	12.1	130		−0.2	−3.3	
	9	1.3	11	484	−1.0	−2.1	+1
11	16.4	13.0	198		+1.2	−3.6	−1
	12	−1.2	−13		+2.0	+0.7	
	8	+4.0	30		+2.0	+7.7	+4
	16.9	13.2	207		−5.3	+1.1	−2
	7	+0.1	1	537	+2.0	−1.2	−1
12	20.8	19.4	374		−4.4	+6.4	−8
	16	−6.0	−89				
Pris'l* for 11	16	−1.4	−119		−4.0	−0.2	
	16	−4.6			−8.0	−8.6	+21
	15.1	14.8	206				
Pris'l* for 11	15.1	15.0			+1.8	+1.8	+1
	12	−0.3		676	−5.0	−0.4	+1
13	15.4	15.7	224		+5.4	−3.7	−6
	12	1.2	13		+4.0	+7.2	+9
	13.2	16.6	203	693	+1.9	+1.8	+1
+47	13.4	12.0	148		+2.0	−3.7	−1
	10	−3.4	−32		+2.0	−4.6	−1
	8	10.0	74	184	+5.2	−6.6	−5
14	12.2	8.0	90		+1.2	−4.0	−1
	4	−4 2	−16		+6.0	−0.8	−1
	8	8.0	59	45	0.0	−2.0	
+20	8	8.0			+4.2	0.0	
Pris'l* for 14	8	0.0			−4.0	+4.2	−1
	8	8.0		3	0.0	0.0	

Approx. volume	+ 2622	+ 40
Prism'l. Cor.	= + 12	− 28
True volume	= 2634	

28. **Rules for Cross Sectioning.**—In deriving the prismoidal formula, the ground surface of each prismoid was assumed to be a hyperbolic paraboloid. Hence if the prismoidal formula, or its equivalent, is to be used :

* These can usually be avoided in practice by placing the same number of breaks in adjacent sections as already indicated in § 20.

Place cross sections at the regular stations (usually 100 feet apart), treating the ground as three-level if the surface is very regular; five-level if at all broken, as the increased labor in field and office is slight; and irregular if more broken. The object being to obtain points enough in each cross section so that *straight lines joining them will lie in the surface, or will equalize the surface,* and to adopt the method which will secure this at most of the sections without extra breaks, even if it gives more points than are necessary at some of the more regular stations.

An occasional extra break can often best be treated by itself, as adding or subtracting a computed amount to or from the regular quantity which would be obtained without it.

It should also be noted that the formula requires the same number of sides or points in adjacent cross sections (dividing the solid into triangular prismoids), requiring that when a ridge or hollow runs out its vanishing point in the more regular section be taken and the section be made up with the break for the solid on one side and without it for the solid on the other side of the section. The height of this vanishing point can be computed, but it is usually better to measure it in the field.

Intermediate cross sections should be added whenever straight lines joining corresponding points of the regular sections will not lie in the surface or equalize the surface, and enough should be added to fulfill this condition. When the center line is curved, the lines joining corresponding points of adjacent cross sections should be estimated at the same curvature in judging of their coincidence with the ground surface.

A cross section should also be placed wherever a grade point occurs on either edge of the roadbed (usually limited to the edges in cutting) and one is often placed where a grade point occurs on the center line. In light work and where the transverse slope is small, a section is sometimes taken where the center line comes to grade and the other two sections omitted. See § 15.

The plus stations are preferably placed at 50 feet, at multiples of 10, or at whole feet (except grade points) from the last full station, for convenience of computing grades in the field and quantities in the office. Breaks on a cross section should be put at multiples of 10, or at whole feet, from the center, when possible without a sacrifice of accuracy.

When the method by averaging end areas is used, without the prismoidal correction, additional sections should be taken whenever the change in transverse slope or in area between adjacent sections is great. The reason for this can be seen by referring to (2); halving the distance between sections will halve both factors $(c - c')$ and $(w' - w)$, or quarter the product, so that when the two half length solids are added together the correction or error will be reduced one-half. The error due to averaging end areas can thus be made as small as desirable, but the method if extended far would soon involve more extra labor than to apply the prismoidal correction.

The same remarks will apply to the method by middle areas.

TABLE

— OF —

Triangular Prismoids for Railway Earthwork

IN CUBIC YARDS,

BY AVERAGING END AREAS.

(See Prismoidal Correction.)

Correction for tenths of height.

.1	.2	.3	.4	.5	.6	.7	.8	.9	WIDTH.	1	2	3	4	5	6	7	8	9	10
					1	1	1	1	1	1	2	3	4	5	6	6	7	8	9
		1	1	1	1	1	1	2	2	2	4	6	7	9	11	13	15	17	19
	1	1	1	1	2	2	2	3	3	3	6	8	11	14	17	19	22	25	28
	1	1	1	2	2	3	3	3	4	4	7	11	15	19	22	26	30	33	37
	1	1	2	2	3	3	4	4	5	5	9	14	19	23	28	32	37	42	46
1	1	2	2	3	3	4	4	5	6	6	11	17	22	28	33	39	44	50	56
1	1	2	3	3	4	5	5	6	7	6	13	19	26	32	39	45	52	58	65
1	1	2	3	4	4	5	6	7	8	7	15	22	30	37	44	52	59	67	74
1	2	3	3	4	5	6	7	8	9	8	17	25	33	42	50	58	67	75	83
1	2	3	4	5	6	6	7	8	10	9	19	28	37	46	56	65	74	83	93
1	2	3	4	5	6	7	8	9	11	10	20	31	41	51	61	71	81	92	102
1	2	3	4	6	7	8	9	10	12	11	22	33	44	56	67	78	89	100	111
1	2	4	5	6	7	8	10	11	13	12	24	36	48	60	72	84	96	108	120
1	3	4	5	6	8	9	10	12	14	13	26	39	52	65	78	91	104	117	130
1	3	4	6	7	8	10	11	13	15	14	28	42	56	69	83	97	111	125	139
1	3	4	6	7	9	10	12	13	16	15	30	44	59	74	89	104	119	133	148
2	3	5	6	8	9	11	13	14	17	16	31	47	63	79	94	110	126	142	157
2	3	5	7	8	10	12	13	15	18	17	33	50	67	83	100	117	133	150	167
2	4	5	7	9	11	12	14	16	19	18	35	53	70	88	106	123	141	158	176
2	4	6	7	9	11	13	15	17	20	19	37	56	74	93	111	130	148	167	185
2	4	6	8	10	12	14	16	18	21	19	39	58	78	97	117	136	156	175	194
2	4	6	8	10	12	14	16	18	22	20	41	61	81	102	122	143	163	183	204
2	4	6	9	11	13	15	17	19	23	21	43	64	85	106	128	149	170	192	213
2	4	7	9	11	13	16	18	20	24	22	44	67	89	111	133	156	178	200	222
2	5	7	9	12	14	16	19	21	25	23	46	69	93	116	139	162	185	208	231
2	5	7	10	12	14	17	19	22	26	24	48	72	96	120	144	169	193	217	241
3	5	8	10	13	15	18	20	23	27	25	50	75	100	125	150	175	200	225	250
3	5	8	10	13	16	18	21	23	28	26	52	78	104	130	156	181	207	233	259
3	5	8	11	13	16	19	21	24	29	27	54	81	107	134	161	188	215	242	269
3	6	8	11	14	17	19	22	25	30	28	56	83	111	139	167	194	222	250	278
3	6	9	11	14	17	20	23	26	31	29	57	86	115	144	172	201	230	258	287
3	6	9	12	15	18	21	24	27	32	30	59	89	119	148	178	207	237	267	296
3	6	9	12	15	18	21	24	28	33	31	61	92	122	153	183	214	244	275	306
3	6	9	13	16	19	22	25	28	34	31	63	94	126	157	189	220	252	283	315
3	6	10	13	16	19	23	26	29	35	32	65	97	130	162	194	227	259	292	324
3	7	10	13	17	20	23	27	30	36	33	67	100	133	167	200	233	267	300	333
3	7	10	14	17	21	24	27	31	37	34	69	103	137	171	206	240	274	308	343
4	7	11	14	18	21	25	28	32	38	35	70	106	141	176	211	246	281	317	352
4	7	11	14	18	22	25	29	33	39	36	72	108	144	181	217	253	289	325	361
4	7	11	15	19	22	26	30	33	40	37	74	111	148	185	222	259	296	333	370
4	8	11	15	19	23	27	30	34	41	38	76	114	152	190	228	266	304	342	380
4	8	12	16	19	23	27	31	35	42	39	78	117	156	194	233	272	311	350	389
4	8	12	16	20	24	28	32	36	43	40	80	119	159	199	239	279	319	358	398
4	8	12	16	20	24	29	33	37	44	41	81	122	163	204	244	285	326	367	407
4	8	13	17	21	25	29	33	38	45	42	83	125	167	208	250	292	333	375	417
4	9	13	17	21	26	30	34	38	46	43	85	128	170	213	256	298	341	383	426
4	9	13	17	22	26	30	35	39	47	44	87	131	174	218	261	305	348	392	435
4	9	13	18	22	27	31	36	40	48	44	89	133	178	222	267	311	356	400	444
5	9	14	18	23	27	32	36	41	49	45	91	136	181	227	272	318	363	408	454
5	9	14	19	23	28	32	37	42	50	46	93	139	185	231	278	324	370	417	463

| .1 | .2 | .3 | .4 | .5 | .6 | .7 | .8 | .9 | | 1 | 2 | 3 | 4 | 5 | 6 | 7 | 8 | 9 | 10 |

Correction for tenths of width.

	1	2	3	4	5	6	7	8	9	10
0.1						1	1	1	1	1
0.2			1	1	1	1	1	1	2	2
0.3		1	1	1	1	2	2	2	3	3
0.4		1	1	1	2	2	3	3	3	4
0.5		1	1	2	2	3	3	4	4	5
0.6	1	1	2	2	3	3	4	4	5	6
0.7	1	1	2	3	3	4	5	5	6	6
0.8	1	1	2	3	4	4	5	6	7	7
0.9	1	2	3	3	4	5	6	7	8	8

Correction for tenths of height.

.1	.2	.3	.4	.5	.6	.7	.8	.9	WIDTH	1	2	3	4	5	6	7	8	9	10
5	9	14	19	24	28	33	38	43	51	47	94	142	189	236	283	331	378	425	472
5	10	14	19	24	29	34	39	43	52	48	96	144	193	241	289	337	385	433	481
5	10	15	20	25	29	34	39	44	53	49	98	147	196	245	294	344	393	442	491
5	10	15	20	25	30	35	40	45	54	50	100	150	200	250	300	350	400	450	500
5	10	15	20	25	31	36	41	46	55	51	102	153	204	255	306	356	407	458	509
5	10	16	21	26	31	36	41	47	56	52	104	156	207	259	311	363	415	467	519
5	11	16	21	26	32	37	42	48	57	53	106	158	211	264	317	369	422	475	528
5	11	16	21	27	32	38	43	48	58	54	107	161	215	269	322	376	430	483	537
5	11	16	22	27	33	38	44	49	59	55	109	164	219	273	328	382	437	492	546
6	11	17	22	28	33	39	44	50	60	56	111	167	222	278	333	389	444	500	556
6	11	17	23	28	34	40	45	51	61	56	113	169	226	282	339	395	452	508	565
6	11	17	23	29	34	40	46	52	62	57	115	172	230	287	344	402	459	517	574
6	12	18	23	29	35	41	47	53	63	58	117	175	233	292	350	408	467	525	583
6	12	18	24	30	36	41	47	53	64	59	119	178	237	296	356	415	474	533	593
6	12	18	24	30	36	42	48	54	65	60	120	181	241	301	361	421	481	542	602
6	12	18	24	31	37	43	49	55	66	61	122	183	244	306	367	428	489	550	611
6	12	19	25	31	37	43	50	56	67	62	124	186	248	310	372	434	496	558	620
6	13	19	25	31	38	44	50	57	68	63	126	189	252	315	378	441	504	567	630
6	13	19	26	32	38	45	51	58	69	64	128	192	256	319	383	447	511	575	639
6	13	19	26	32	39	45	52	58	70	65	130	194	259	324	389	454	519	583	648
7	13	20	26	33	39	46	53	59	71	66	131	197	263	329	394	460	526	592	657
7	13	20	27	33	40	47	53	60	72	67	133	200	267	333	400	467	533	600	667
7	14	20	27	34	41	47	54	61	73	68	135	203	270	338	406	473	541	608	676
7	14	21	27	34	41	48	55	62	74	69	137	206	274	343	411	480	548	617	685
7	14	21	28	35	42	49	56	63	75	69	139	208	278	347	417	486	556	625	694
7	14	21	28	35	42	49	56	63	76	70	141	211	281	352	422	493	563	633	704
7	14	21	29	36	43	50	57	64	77	71	143	214	285	356	428	499	570	642	713
7	14	22	29	36	43	51	58	65	78	72	144	217	289	361	433	506	578	650	722
7	15	22	29	37	44	51	59	66	79	73	146	219	293	366	439	512	585	658	731
7	15	22	30	37	44	52	59	67	80	74	148	222	296	370	444	519	593	667	741
8	15	23	30	37	45	53	60	68	81	75	150	225	300	375	450	525	600	675	750
8	15	23	30	38	46	53	61	68	82	76	152	228	304	380	456	531	607	683	759
8	15	23	31	38	46	54	61	69	83	77	154	231	307	384	461	538	615	692	769
8	16	23	31	39	47	54	62	70	84	78	156	233	311	389	467	544	622	700	778
8	16	24	31	39	47	55	63	71	85	79	157	236	315	394	472	551	630	708	787
8	16	24	32	40	48	56	64	72	86	80	159	239	319	398	478	557	637	717	796
8	16	24	32	40	48	56	64	73	87	81	161	242	322	403	483	564	644	725	806
8	16	24	33	41	49	57	65	73	88	81	163	244	326	407	489	570	652	733	815
8	16	25	33	41	49	58	66	74	89	82	165	247	330	412	494	577	659	742	824
8	17	25	33	42	50	58	67	75	90	83	167	250	333	417	500	583	667	750	833
8	17	25	34	42	51	59	67	76	91	84	169	253	337	421	506	590	674	758	843
9	17	26	34	43	51	60	68	77	92	85	170	256	341	426	511	596	681	767	852
9	17	26	34	43	52	60	69	78	93	86	172	258	344	431	517	603	689	775	861
9	17	26	35	44	52	61	70	78	94	87	174	261	348	435	522	609	696	783	870
9	18	26	35	44	53	62	70	79	95	88	176	264	352	440	528	616	704	792	880
9	18	27	36	44	53	62	71	80	96	89	178	267	356	444	533	622	711	800	889
9	18	27	36	45	54	63	72	81	97	90	180	269	359	449	539	629	719	808	898
9	18	27	36	45	54	64	73	82	98	91	181	272	363	454	544	635	726	817	907
9	18	28	37	46	55	64	73	83	99	92	183	275	367	458	550	642	733	825	917
9	19	28	37	46	56	65	74	83	100	93	185	278	370	463	556	648	741	833	926

Correction for tenths of width.

	1	2	3	4	5	6	7	8	9	10
0.1						1	1	1	1	1
0.2			1	1	1	1	1	2	2	2
0.3		1	1	1	2	2	2	2	3	3
0.4		1	1	2	2	2	3	3	4	4
0.5	1	1	2	2	3	3	4	4	5	5
0.6	1	1	2	2	3	3	4	4	5	6
0.7	1	1	2	3	3	4	5	5	6	6
0.8	1	1	2	3	3	4	4	5	6	7
0.9	1	2	3	3	4	5	6	7	8	8

Correction for tenths of height.

.1	.2	.3	.4	.5	.6	.7	.8	.9	WIDTH.	21	22	23	24	25	26	27	28	29	30
					1	1	1	1	1	19	20	21	22	23	24	25	26	27	28
		1	1	1	1	1	1	2	2	39	41	43	44	46	48	50	52	54	56
	1	1	1	2	2	2	2	3	3	58	61	64	67	69	72	75	78	81	83
	1	1	1	2	2	3	3	3	4	78	81	85	89	93	96	100	104	107	111
	1	1	2	2	3	3	4	4	5	97	102	106	111	116	120	125	130	134	139
1	1	2	2	3	3	4	4	5	6	117	122	128	133	139	144	150	156	161	167
1	1	2	3	3	4	5	5	6	7	136	143	149	156	162	169	175	181	188	194
1	1	2	3	4	4	5	6	7	8	156	163	170	178	185	193	200	207	215	222
1	2	3	3	4	5	6	7	8	9	175	183	192	200	208	217	225	233	242	250
1	2	3	4	5	6	6	7	8	10	194	204	213	222	231	241	250	259	269	278
1	2	3	4	5	6	7	8	9	11	214	224	234	244	255	265	275	285	295	306
1	2	3	4	6	7	8	9	10	12	233	244	256	267	278	289	300	311	322	333
1	2	4	5	6	7	8	10	11	13	253	265	277	289	301	313	325	337	349	361
1	3	4	5	6	8	9	10	12	14	272	285	298	311	324	337	350	363	376	389
1	3	4	6	7	8	10	11	12	15	292	306	319	333	347	361	375	389	403	417
1	3	4	6	7	9	10	12	13	16	311	326	341	356	370	385	400	415	430	444
2	3	5	6	8	9	11	13	14	17	331	346	362	378	394	409	425	441	456	472
2	3	5	7	8	10	12	13	15	18	350	367	383	400	417	433	450	467	483	500
2	4	5	7	9	11	12	14	16	19	369	387	405	422	440	457	475	493	510	528
2	4	6	7	9	11	13	15	17	20	389	407	426	444	463	481	500	519	537	556
2	4	6	8	10	12	14	16	18	21	408	428	447	467	486	506	525	544	564	583
2	4	6	8	10	12	14	16	18	22	428	448	469	489	509	530	550	570	591	611
2	4	6	9	11	13	15	17	19	23	447	469	490	511	532	554	575	596	618	639
2	4	7	9	11	13	16	18	20	24	467	489	511	533	556	578	600	622	644	667
2	5	7	9	12	14	16	19	21	25	486	509	532	556	579	602	625	648	671	694
2	5	7	10	12	14	17	19	22	26	506	530	554	578	602	626	650	674	698	722
3	5	8	10	12	15	18	20	23	27	525	550	575	600	625	650	675	700	725	750
3	5	8	10	13	16	18	21	23	28	544	570	596	622	648	674	700	726	752	778
3	5	8	11	13	16	19	21	24	29	564	591	618	644	671	698	725	752	779	806
3	6	8	11	14	17	19	22	25	30	583	611	639	667	694	722	750	778	806	833
3	6	9	11	14	17	20	23	26	31	603	631	660	689	718	746	775	804	832	861
3	6	9	12	15	18	21	24	27	32	622	652	681	711	741	770	800	830	859	889
3	6	9	12	15	18	21	24	28	33	642	672	703	733	764	794	825	856	886	917
3	6	9	13	16	19	22	25	28	34	661	693	724	756	787	819	850	881	913	944
3	6	10	13	16	19	23	26	29	35	681	713	745	778	810	843	875	907	940	972
3	7	10	13	17	20	23	27	30	36	700	733	767	800	833	867	900	933	967	1000
3	7	10	14	17	21	24	27	31	37	719	754	788	822	856	891	925	959	994	1028
4	7	11	14	18	21	25	28	32	38	739	774	809	844	880	915	950	985	1020	1056
4	7	11	14	18	22	25	29	33	39	758	794	831	867	903	939	975	1011	1047	1083
4	7	11	15	19	22	26	30	33	40	778	815	852	889	926	963	1000	1037	1074	1111
4	8	11	15	19	23	27	30	34	41	797	835	873	911	949	987	1025	1063	1101	1139
4	8	12	16	19	23	27	31	35	42	817	856	894	933	972	1011	1050	1089	1128	1167
4	8	12	16	20	24	28	32	36	43	836	876	916	956	995	1035	1075	1115	1155	1194
4	8	12	16	20	24	29	33	37	44	856	896	937	978	1019	1059	1100	1141	1181	1222
4	8	13	17	21	25	29	33	38	45	875	917	958	1000	1042	1083	1125	1167	1208	1250
4	9	13	17	21	26	30	34	38	46	894	937	980	1022	1065	1107	1150	1193	1235	1278
4	9	13	17	22	26	30	35	39	47	914	957	1001	1044	1088	1131	1175	1219	1262	1306
4	9	13	18	22	27	31	36	40	48	933	978	1022	1067	1111	1156	1200	1244	1289	1333
5	9	14	18	23	27	32	36	41	49	953	998	1044	1089	1134	1180	1225	1270	1316	1361
5	9	14	19	23	28	32	37	42	50	972	1019	1065	1111	1157	1204	1250	1296	1343	1389
.1	.2	.3	.4	.5	.6	.7	.8	.9		21	22	23	24	25	26	27	28	29	30

Correction for tenths of width.

				21	22	23	24	25	26	27	28	29	30
0.1				2	2	2	2	2	2	3	3	3	3
0.2				4	4	4	4	5	5	5	5	5	6
0.3				6	6	6	7	7	7	8	8	8	8
0.4				8	8	9	9	9	10	10	10	11	11
0.5				10	10	11	11	12	12	12	13	13	14
0.6				12	12	13	13	14	14	15	16	16	17
0.7				14	14	15	16	16	17	18	18	19	19
0.8				16	16	17	18	19	19	20	21	21	22
0.9				18	18	19	20	21	22	23	23	24	25

.1	.2	.3	.4	.5	.6	.7	.8	.9	WIDTH.	21	22	23	24	25	26	27	28	29	30
5	9	14	19	24	28	33	38	43	51	992	1039	1086	1133	1181	1228	1275	1322	1369	1417
5	10	14	19	24	29	34	39	43	52	1011	1059	1107	1156	1204	1252	1300	1348	1396	1444
5	10	15	20	25	29	34	39	44	53	1031	1080	1129	1178	1227	1276	1325	1374	1423	1472
5	10	15	20	25	30	35	40	45	54	1050	1100	1150	1200	1250	1300	1350	1400	1450	1500
5	10	15	20	25	31	36	41	46	55	1069	1120	1171	1222	1273	1324	1375	1426	1477	1528
5	10	16	21	26	31	36	41	47	56	1089	1141	1193	1244	1296	1348	1400	1452	1504	1556
5	11	16	21	26	32	37	42	48	57	1108	1161	1214	1267	1319	1372	1425	1478	1531	1583
5	11	16	21	27	32	38	43	48	58	1128	1181	1235	1289	1343	1396	1450	1504	1557	1611
5	11	16	22	27	33	38	44	49	59	1147	1202	1256	1311	1366	1420	1475	1530	1584	1639
6	11	17	22	28	33	39	44	50	60	1167	1222	1278	1333	1389	1444	1500	1556	1611	1667
6	11	17	23	28	34	40	45	51	61	1186	1243	1299	1356	1412	1469	1525	1581	1638	1694
6	11	17	23	29	34	40	46	52	62	1206	1263	1320	1378	1435	1493	1550	1607	1665	1722
6	12	18	23	29	35	41	47	53	63	1225	1283	1342	1400	1458	1517	1575	1633	1692	1750
6	12	18	24	30	36	41	47	53	64	1244	1304	1363	1422	1481	1541	1600	1659	1719	1778
6	12	18	24	30	36	42	48	54	65	1264	1324	1384	1444	1505	1565	1625	1685	1745	1806
6	12	18	24	31	37	43	49	55	66	1283	1344	1406	1467	1528	1589	1650	1711	1772	1833
6	12	19	25	31	37	43	50	56	67	1303	1365	1427	1489	1551	1613	1675	1737	1799	1861
6	13	19	25	31	38	44	50	57	68	1322	1385	1448	1511	1574	1637	1700	1763	1826	1889
6	13	19	26	32	38	45	51	58	69	1342	1406	1469	1533	1597	1661	1725	1789	1853	1917
6	13	19	26	32	39	45	52	58	70	1361	1426	1491	1556	1620	1685	1750	1815	1880	1944
7	13	20	26	33	39	46	53	59	71	1381	1446	1512	1578	1644	1709	1775	1841	1906	1972
7	13	20	27	33	40	47	53	60	72	1400	1467	1533	1600	1667	1733	1800	1867	1933	2000
7	14	20	27	34	41	47	54	61	73	1419	1487	1555	1622	1690	1757	1825	1893	1960	2028
7	14	21	27	34	41	48	55	62	74	1439	1507	1576	1644	1713	1781	1850	1919	1987	2056
7	14	21	28	35	42	49	56	63	75	1458	1528	1597	1667	1736	1806	1875	1944	2014	2083
7	14	21	28	35	42	49	56	63	76	1478	1548	1619	1689	1759	1830	1900	1970	2041	2111
7	14	21	29	36	43	50	57	64	77	1497	1569	1640	1711	1782	1854	1925	1996	2068	2139
7	14	22	29	36	43	51	58	65	78	1517	1589	1661	1733	1806	1878	1950	2022	2094	2167
7	15	22	29	37	44	51	59	66	79	1536	1609	1682	1756	1829	1902	1975	2048	2121	2194
7	15	22	30	37	44	52	59	67	80	1556	1630	1704	1778	1852	1926	2000	2074	2148	2222
8	15	23	30	37	45	53	60	68	81	1575	1650	1725	1800	1875	1950	2025	2100	2175	2250
8	15	23	30	38	46	53	61	68	82	1594	1670	1746	1822	1898	1974	2050	2126	2202	2278
8	15	23	31	38	46	54	61	69	83	1614	1691	1768	1844	1921	1998	2075	2152	2229	2306
8	16	23	31	39	47	54	62	70	84	1633	1711	1789	1867	1944	2022	2100	2178	2256	2333
8	16	24	31	39	47	55	63	71	85	1653	1731	1810	1889	1968	2046	2125	2204	2282	2361
8	16	24	32	40	48	56	64	72	86	1672	1752	1831	1911	1991	2070	2150	2230	2309	2389
8	16	24	32	40	48	56	64	73	87	1692	1772	1853	1933	2014	2094	2175	2256	2336	2417
8	16	24	33	41	49	57	65	73	88	1711	1793	1874	1956	2037	2119	2200	2281	2363	2444
8	16	25	33	41	49	58	66	74	89	1731	1813	1895	1978	2060	2143	2225	2307	2390	2472
8	17	25	33	42	50	58	67	75	90	1750	1833	1917	2000	2083	2167	2250	2333	2417	2500
8	17	25	34	42	51	59	67	76	91	1769	1854	1938	2022	2106	2191	2275	2359	2444	2528
9	17	26	34	43	51	60	68	77	92	1789	1874	1959	2044	2130	2215	2300	2385	2470	2556
9	17	26	34	43	52	60	69	78	93	1808	1894	1981	2067	2153	2239	2325	2411	2497	2583
9	17	26	35	44	52	61	70	78	94	1828	1915	2002	2089	2176	2263	2350	2437	2524	2611
9	18	26	35	44	53	62	70	79	95	1847	1935	2023	2111	2199	2287	2375	2463	2551	2639
9	18	27	36	44	53	62	71	80	96	1867	1956	2044	2133	2222	2311	2400	2489	2578	2667
9	18	27	36	45	54	63	72	81	97	1886	1976	2066	2156	2245	2335	2425	2515	2605	2694
9	18	27	36	45	54	64	73	82	98	1906	1996	2087	2178	2269	2359	2450	2541	2631	2722
9	18	28	37	46	55	64	73	83	99	1925	2017	2108	2200	2292	2383	2475	2567	2658	2750
9	19	28	37	46	56	65	74	83	100	1944	2037	2130	2222	2315	2407	2500	2593	2685	2778
.1	.2	.3	.4	.5	.6	.7	.8	.9		21	22	23	24	25	26	27	28	29	30

Correction for tenths of height.

Correction for tenths of width.

	21	22	23	24	25	26	27	28	29	30
0.1	2	2	2	2	2	2	3	3	3	3
0.2	4	4	4	4	5	5	5	5	5	6
0.3	6	6	6	7	7	7	8	8	8	8
0.4	8	8	9	9	9	10	10	10	11	11
0.5	10	10	11	11	12	12	13	13	14	14
0.6	12	12	13	13	14	14	15	16	16	17
0.7	14	14	15	16	16	17	18	18	19	19
0.8	16	16	17	18	18	19	19	20	21	22
0.9	18	18	19	19	20	21	22	23	24	25

Correction for tenths of height. (left margin)

.1	.2	.3	.4	.5	.6	.7	.8	.9	WIDTH.	31	32	33	34	35	36	37	38	39	40
				1	1	1	1	1	1	29	30	31	31	32	33	34	35	36	37
		1	1	1	1	1	1	2	2	57	59	61	63	65	67	69	70	72	74
	1	1	1	1	2	2	2	3	3	86	89	92	94	97	100	103	106	108	111
1	1	1	2	2	3	3	3	3	4	115	119	122	126	130	133	137	141	144	148
1	1	2	2	3	3	4	4		5	144	148	153	157	162	167	171	176	181	185
1	1	2	2	3	3	4	4	5	6	172	178	183	189	194	200	206	211	217	222
1	1	2	3	3	4	5	5	6	7	201	207	214	220	227	233	240	246	253	259
1	1	2	3	4	4	5	6	7	8	230	237	244	252	259	267	274	281	289	296
1	2	3	3	4	5	6	7	8	9	258	267	275	283	292	300	308	317	325	333
1	2	3	4	5	6	6	7	8	10	287	296	306	315	324	333	343	352	361	370
1	2	3	4	5	6	7	8	9	11	316	326	336	346	356	367	377	387	397	407
1	2	3	4	6	7	8	9	10	12	344	356	367	378	389	400	411	422	433	444
1	2	4	5	6	7	8	10	11	13	373	385	397	409	421	433	445	457	469	481
1	3	4	5	6	8	9	10	12	14	402	415	428	441	454	467	480	493	506	519
1	3	4	6	7	8	10	11	12	15	431	444	458	472	486	500	514	528	542	566
1	3	4	6	7	9	10	12	13	16	459	474	489	504	519	533	548	563	578	593
2	3	5	6	8	9	11	13	14	17	488	504	519	535	551	567	582	598	614	630
2	3	5	7	8	10	12	13	15	18	517	533	550	567	583	600	617	633	650	667
2	4	5	7	9	11	12	14	16	19	545	563	581	598	616	633	651	669	686	704
2	4	6	7	9	11	13	15	17	20	574	593	611	630	648	667	685	704	722	741
2	4	6	8	10	12	14	16	18	21	603	622	642	661	681	700	719	739	758	778
2	4	6	8	10	12	14	16	18	22	631	652	672	693	713	733	754	774	794	815
2	4	6	9	11	13	15	17	19	23	660	681	703	724	745	767	788	809	831	852
2	4	7	9	11	13	16	18	20	24	689	711	733	756	778	800	822	844	867	889
2	5	7	9	12	14	16	19	21	25	718	741	764	787	810	833	856	880	903	926
2	5	7	10	12	14	17	19	22	26	746	770	794	819	843	867	891	915	939	963
3	5	8	10	12	15	18	20	23	27	775	800	825	850	875	900	925	950	975	1000
3	5	8	10	13	16	18	21	23	28	804	830	856	881	907	933	959	985	1011	1037
3	5	8	11	13	16	19	21	24	29	832	859	886	913	940	967	994	1020	1047	1074
3	6	8	11	14	17	19	22	25	30	861	889	917	944	972	1000	1028	1056	1083	1111
3	6	9	11	14	17	20	23	26	31	890	919	947	976	1005	1033	1062	1091	1119	1148
3	6	9	12	15	18	21	24	27	32	919	948	978	1007	1037	1067	1096	1126	1156	1185
3	6	9	12	15	18	21	24	28	33	947	978	1008	1039	1069	1100	1131	1161	1192	1222
3	6	9	13	16	19	22	25	28	34	976	1007	1039	1070	1102	1133	1165	1196	1228	1259
3	6	10	13	16	19	23	26	29	35	1005	1037	1069	1102	1134	1167	1199	1231	1264	1296
3	7	10	13	17	20	23	27	30	36	1033	1067	1100	1133	1167	1200	1233	1267	1300	1333
3	7	10	14	17	21	24	27	31	37	1062	1096	1131	1165	1199	1233	1268	1302	1336	1370
4	7	11	14	18	21	25	28	32	38	1091	1126	1161	1196	1231	1267	1302	1337	1372	1407
4	7	11	14	18	22	25	29	33	39	1119	1156	1192	1228	1264	1300	1336	1372	1408	1444
4	7	11	15	19	22	26	30	33	40	1148	1185	1222	1259	1296	1333	1370	1407	1444	1481
4	8	11	15	19	23	27	30	34	41	1177	1215	1253	1291	1329	1367	1405	1443	1481	1519
4	8	12	16	19	23	27	31	35	42	1206	1244	1283	1322	1361	1400	1439	1478	1517	1556
4	8	12	16	20	24	28	32	36	43	1234	1274	1314	1354	1394	1433	1473	1513	1553	1593
4	8	12	16	20	24	29	33	37	44	1263	1304	1344	1385	1426	1467	1507	1548	1589	1630
4	8	13	17	21	25	29	33	38	45	1292	1333	1375	1417	1458	1500	1542	1583	1625	1667
4	9	13	17	21	26	30	34	38	46	1320	1363	1406	1448	1491	1533	1576	1619	1661	1704
4	9	13	17	22	26	30	35	39	47	1349	1393	1436	1480	1523	1567	1610	1654	1697	1741
4	9	13	18	22	27	31	36	40	48	1378	1422	1467	1511	1556	1600	1644	1689	1733	1778
5	9	14	18	23	27	32	36	41	49	1406	1452	1497	1543	1588	1633	1679	1724	1769	1815
5	9	14	19	23	28	32	37	42	50	1435	1481	1528	1574	1620	1667	1713	1759	1806	1852
.1	.2	.3	.4	.5	.6	.7	.8	.9		31	32	33	34	35	36	37	38	39	40

Correction for tenths of width.

	31	32	33	34	35	36	37	38	39	40
0.1	3	3	3	3	3	3	3	4	4	4
0.2	6	6	6	6	6	7	7	7	7	7
0.3	9	9	9	9	10	10	10	11	11	11
0.4	11	12	12	13	13	13	14	14	14	15
0.5	14	15	15	16	16	17	17	18	18	19
0.6	17	18	18	19	19	20	21	21	22	22
0.7	20	21	21	22	23	23	24	25	25	26
0.8	23	24	24	25	26	27	27	28	29	30
0.9	26	27	28	28	29	30	31	32	33	33

.1	.2	.3	.4	.5	.6	.7	.8	.9	WIDTH.	31	32	33	34	35	36	37	38	39	40
5	9	14	19	24	28	33	38	43	51	1464	1511	1558	1606	1653	1700	1747	1794	1842	1889
5	10	14	19	24	29	34	39	43	52	1493	1541	1589	1637	1685	1733	1781	1830	1878	1926
5	10	15	20	25	29	34	39	44	53	1521	1570	1619	1669	1718	1767	1816	1865	1914	1963
5	10	15	20	25	30	35	40	45	54	1550	1600	1650	1700	1750	1800	1850	1900	1950	2000
5	10	15	20	25	31	36	41	46	55	1579	1630	1681	1731	1782	1833	1884	1935	1986	2037
5	10	16	21	26	31	36	41	47	56	1607	1659	1711	1763	1815	1867	1919	1970	2022	2074
5	11	16	21	26	32	37	42	48	57	1636	1689	1742	1794	1847	1900	1953	2006	2058	2111
5	11	16	21	27	32	38	43	48	58	1665	1719	1772	1826	1880	1933	1987	2041	2094	2148
5	11	16	22	27	33	38	44	49	59	1694	1748	1803	1857	1912	1967	2021	2076	2131	2185
6	11	17	22	28	33	39	44	50	60	1722	1778	1833	1889	1944	2000	2056	2111	2167	2222
6	11	17	23	28	34	40	45	51	61	1751	1807	1864	1920	1977	2033	2090	2146	2203	2259
6	11	17	23	29	34	40	46	52	62	1780	1837	1894	1952	2009	2067	2124	2181	2239	2296
6	12	18	23	29	35	41	47	53	63	1808	1867	1925	1983	2042	2100	2158	2217	2275	2333
6	12	18	24	30	36	41	47	53	64	1837	1896	1956	2015	2074	2133	2193	2252	2311	2370
6	12	18	24	30	36	42	48	54	65	1866	1926	1986	2046	2106	2167	2227	2287	2347	2407
6	12	18	24	31	37	43	49	55	66	1894	1956	2017	2078	2139	2200	2261	2322	2383	2444
6	12	19	25	31	37	43	50	56	67	1923	1985	2047	2109	2171	2233	2295	2357	2419	2481
6	13	19	25	31	38	44	50	57	68	1952	2015	2078	2141	2204	2267	2330	2393	2456	2519
6	13	19	26	32	38	45	51	58	69	1981	2044	2108	2172	2236	2300	2364	2428	2492	2556
6	13	19	26	32	39	45	52	58	70	2009	2074	2139	2204	2269	2333	2398	2463	2528	2593
7	13	20	26	33	39	46	53	59	71	2038	2104	2169	2235	2301	2367	2432	2498	2564	2630
7	13	20	27	33	40	47	53	60	72	2067	2133	2200	2267	2333	2400	2467	2533	2600	2667
7	14	20	27	34	41	47	54	61	73	2095	2163	2231	2298	2366	2433	2501	2569	2636	2704
7	14	21	27	34	41	48	55	62	74	2124	2193	2261	2330	2398	2467	2535	2604	2672	2741
7	14	21	28	35	42	49	56	63	75	2153	2222	2292	2361	2431	2500	2569	2639	2708	2778
7	14	21	28	35	42	49	56	63	76	2181	2252	2322	2393	2463	2533	2604	2674	2744	2815
7	14	21	29	36	43	50	57	64	77	2210	2281	2353	2424	2495	2567	2638	2709	2781	2852
7	14	22	29	36	43	51	58	65	78	2239	2311	2383	2456	2528	2600	2672	2744	2817	2889
7	15	22	29	37	44	51	59	66	79	2268	2341	2414	2487	2560	2633	2706	2780	2853	2926
7	15	22	30	37	44	52	59	67	80	2296	2370	2444	2519	2593	2667	2741	2815	2889	2963
8	15	23	30	37	45	53	60	68	81	2325	2400	2475	2550	2625	2700	2775	2850	2925	3000
8	15	23	30	38	46	53	61	68	82	2354	2430	2506	2581	2657	2733	2809	2885	2961	3037
8	15	23	31	38	46	54	61	69	83	2382	2459	2536	2613	2690	2767	2844	2920	2997	3074
8	16	23	31	39	47	54	62	70	84	2411	2489	2567	2644	2722	2800	2878	2956	3033	3111
8	16	24	31	39	47	55	63	71	85	2440	2519	2597	2676	2755	2833	2912	2991	3069	3148
8	16	24	32	40	48	56	64	72	86	2469	2548	2628	2707	2787	2867	2946	3026	3106	3185
8	16	24	32	40	48	56	64	73	87	2497	2578	2658	2739	2819	2900	2981	3061	3142	3222
8	16	24	33	41	49	57	65	73	88	2526	2607	2689	2770	2852	2933	3015	3096	3178	3259
8	16	25	33	41	49	58	66	74	89	2555	2637	2719	2802	2884	2967	3049	3131	3214	3296
8	17	25	33	42	50	58	67	75	90	2583	2667	2750	2833	2917	3000	3083	3167	3250	3333
8	17	25	34	42	51	59	67	76	91	2612	2696	2781	2865	2949	3033	3118	3202	3286	3370
9	17	26	34	43	51	60	68	77	92	2641	2726	2811	2896	2981	3067	3152	3237	3322	3407
9	17	26	34	43	52	60	69	78	93	2669	2756	2842	2928	3014	3100	3186	3272	3358	3444
9	17	26	35	44	52	61	70	78	94	2698	2785	2872	2959	3046	3133	3220	3307	3394	3481
9	18	26	35	44	53	62	70	79	95	2727	2815	2903	2991	3079	3167	3255	3343	3431	3519
9	18	27	36	44	53	62	71	80	96	2756	2844	2933	3022	3111	3200	3289	3378	3467	3556
9	18	27	36	45	54	63	72	81	97	2784	2874	2964	3054	3144	3233	3323	3413	3503	3593
9	18	27	36	45	54	64	73	82	98	2813	2904	2994	3085	3176	3267	3357	3448	3539	3630
9	18	28	37	46	55	64	73	83	99	2842	2933	3025	3117	3208	3300	3392	3483	3575	3667
9	19	28	37	46	56	65	74	83	100	2870	2963	3056	3148	3241	3333	3426	3519	3611	3704

.1	.2	.3	.4	.5	.6	.7	.8	.9		31	32	33	34	35	36	37	38	39	40

		31	32	33	34	35	36	37	38	39	40
	0.1	3	3	3	3	3	3	3	4	4	4
	0.2	6	6	6	6	6	6	7	7	7	7
	0.3	9	9	9	9	10	10	10	11	11	11
Correction for tenths	0.4	11	12	12	13	13	13	14	14	14	15
	0.5	14	15	15	16	16	17	17	18	18	19
of width.	0.6	17	18	18	19	19	20	21	21	22	22
	0.7	20	21	21	22	23	23	24	25	25	26
	0.8	23	24	24	25	26	27	27	28	29	30
	0.9	26	27	28	28	29	30	31	32	33	33

Correction for tenths of height.

.1	.2	.3	.4	.5	.6	.7	.8	.9	WIDTH.	41	42	43	44	45	46	47	48	49	50
					1	1	1	1	1	38	39	40	41	42	43	44	44	45	46
	1	1	1	1	1	1	1	2	2	76	78	80	81	83	85	87	89	91	93
1	1	1	1	2	2	2	2	3	3	114	117	119	122	125	128	131	133	136	139
1	1	1	2	2	3	3	3	3	4	152	156	159	163	167	170	174	178	181	185
1	1	2	2	3	3	3	4	4	5	190	194	199	204	208	213	218	222	227	231
1	1	2	2	3	4	4	4	5	6	228	233	239	244	250	256	261	267	272	278
1	1	2	3	3	4	5	5	6	7	266	272	279	285	292	298	305	311	318	324
1	1	2	3	4	4	5	6	7	8	304	311	319	326	333	341	348	356	363	370
1	2	3	3	4	5	6	7	8	·9	342	350	358	367	375	383	392	400	408	417
1	2	3	4	5	6	6	7	8	10	380	389	398	407	417	426	435	444	454	463
1	2	3	4	5	6	7	8	9	11	418	428	438	448	458	469	479	489	499	509
1	2	3	4	6	7	8	9	10	12	456	467	478	489	500	511	522	533	544	556
1	2	4	5	6	7	8	10	11	13	494	506	518	530	542	554	566	578	590	602
1	3	4	5	6	8	9	10	12	14	531	544	557	570	583	596	609	622	635	648
1	3	4	6	7	8	10	11	12	15	569	583	597	611	625	639	653	667	681	694
1	3	4	6	7	9	10	12	13	16	607	622	637	652	667	681	696	711	726	741
2	3	5	6	8	9	11	13	14	17	645	661	677	693	708	724	740	756	771	787
2	3	5	7	8	10	12	13	15	18	683	700	717	733	750	767	783	800	817	833
2	4	5	7	9	11	12	14	16	19	721	739	756	774	792	809	827	844	862	880
2	4	6	7	9	11	13	15	17	20	759	778	796	815	833	852	870	889	907	926
2	4	6	8	10	12	14	16	18	21	797	817	836	856	875	894	914	933	953	972
2	4	6	8	10	12	14	16	18	22	835	856	876	896	917	937	957	978	998	1019
2	4	6	9	11	13	15	17	19	23	873	894	916	937	958	980	1001	1022	1044	1065
2	4	7	9	11	13	16	18	20	24	911	933	956	978	1000	1022	1044	1067	1089	1111
2	5	7	9	12	14	16	19	21	25	949	972	995	1019	1042	1065	1088	1111	1134	1157
2	5	7	10	12	14	17	19	22	26	987	1011	1035	1059	1083	1107	1131	1156	1180	1204
3	5	8	10	12	15	18	20	23	27	1025	1050	1075	1100	1125	1150	1175	1200	1225	1250
3	5	8	10	13	16	18	21	23	28	1063	1089	1115	1141	1167	1193	1219	1244	1270	1296
3	5	8	11	13	16	19	21	24	29	1101	1128	1155	1181	1208	1235	1262	1289	1316	1343
3	6	8	11	14	17	19	22	25	30	1139	1167	1194	1222	1250	1278	1306	1333	1361	1389
3	6	9	11	14	17	20	23	26	31	1177	1206	1234	1263	1292	1320	1349	1378	1406	1435
3	6	9	12	15	18	21	24	27	32	1215	1244	1274	1304	1333	1363	1393	1422	1452	1481
3	6	9	12	15	18	21	24	28	33	1253	1283	1314	1344	1375	1406	1436	1467	1497	1528
3	6	9	13	16	19	22	25	28	34	1291	1322	1354	1385	1417	1448	1480	1511	1543	1574
3	6	10	13	16	19	23	26	29	35	1329	1361	1394	1426	1458	1491	1523	1556	1588	1620
3	7	10	13	17	20	23	27	30	36	1367	1400	1433	1467	1500	1533	1567	1600	1633	1667
3	7	10	14	17	21	24	27	31	37	1405	1439	1473	1507	1542	1576	1610	1644	1679	1713
4	7	11	14	18	21	25	28	32	38	1443	1478	1513	1548	1583	1619	1654	1689	1724	1759
4	7	11	14	18	22	25	29	33	39	1481	1517	1553	1589	1625	1661	1697	1733	1769	1806
4	7	11	15	19	22	26	30	33	40	1519	1556	1593	1630	1667	1704	1741	1778	1815	1852
4	8	11	15	19	23	27	30	34	41	1556	1594	1632	1670	1708	1746	1784	1822	1860	1898
4	8	12	16	19	23	27	31	35	42	1594	1633	1672	1711	1750	1789	1828	1867	1906	1944
4	8	12	16	20	24	28	32	36	43	1632	1672	1712	1752	1792	1831	1871	1911	1951	1991
4	8	12	16	20	24	29	33	37	44	1670	1711	1752	1793	1833	1874	1915	1956	1996	2037
4	8	13	17	21	25	29	33	38	45	1708	1750	1792	1833	1875	1917	1958	2000	2042	2083
4	9	13	17	21	26	30	34	38	46	1746	1789	1831	1874	1917	1959	2002	2044	2087	2130
4	9	13	17	22	26	30	35	39	47	1784	1828	1871	1915	1958	2002	2045	2089	2132	2176
4	9	13	18	22	27	31	36	40	48	1822	1867	1911	1956	2000	2044	2089	2133	2178	2222
5	9	14	18	23	27	32	36	41	49	1860	1906	1951	1996	2042	2087	2132	2178	2223	2269
5	9	14	19	23	28	32	37	42	50	1898	1944	1991	2037	2083	2130	2176	2222	2269	2315

.1	.2	.3	.4	.5	.6	.7	.8	.9		41	42	43	44	45	46	47	48	49	50

Correction for tenths of width.

	41	42	43	44	45	46	47	48	49	50
0.1	4	4	4	4	4	4	4	4	5	5
0.2	8	8	8	8	8	9	9	9	9	9
0.3	11	12	12	12	13	13	13	13	14	14
0.4	15	16	16	16	17	17	17	18	18	19
0.5	19	19	20	20	21	21	22	22	23	23
0.6	23	23	24	24	25	26	26	27	27	28
0.7	27	27	28	29	29	30	30	31	32	32
0.8	30	31	32	33	33	34	35	36	36	37
0.9	34	35	36	37	38	38	39	40	41	42

Correction for tenths of height.

END AREAS.

.1	.2	.3	.4	.5	.6	.7	.8	.9	WIDTH.	41	42	43	44	45	46	47	48	49	50
5	9	14	19	24	28	33	38	43	51	1936	1983	2031	2078	2125	2172	2219	2267	2314	2361
5	10	14	19	24	29	34	39	43	52	1974	2022	2070	2119	2167	2215	2263	2311	2359	2407
5	10	15	20	25	29	34	39	44	53	2012	2061	2110	2159	2208	2257	2306	2356	2405	2454
5	10	15	20	25	30	35	40	45	54	2050	2100	2150	2200	2250	2300	2350	2400	2450	2500
5	10	15	20	25	31	36	41	46	55	2088	2139	2190	2241	2292	2343	2394	2444	2495	2546
5	10	16	21	26	31	36	41	47	56	2126	2178	2230	2281	2333	2385	2437	2489	2541	2593
5	11	16	21	26	32	37	42	48	57	2164	2217	2269	2322	2375	2428	2481	2533	2586	2639
5	11	16	21	27	32	38	43	48	58	2202	2256	2309	2363	2417	2470	2524	2578	2631	2685
5	11	16	22	27	33	38	44	49	59	2240	2294	2349	2404	2458	2513	2568	2622	2677	2731
6	11	17	22	28	33	39	44	50	60	2278	2333	2389	2444	2500	2556	2611	2667	2722	2778
6	11	17	23	28	34	40	45	51	61	2316	2372	2429	2485	2542	2598	2655	2711	2768	2824
6	11	17	23	29	34	40	46	52	62	2354	2411	2469	2526	2583	2641	2698	2756	2813	2870
6	12	18	23	29	35	41	47	53	63	2392	2450	2508	2567	2625	2683	2742	2800	2858	2917
6	12	18	24	30	36	41	47	53	64	2430	2489	2548	2607	2667	2726	2785	2844	2904	2963
6	12	18	24	30	36	42	48	54	65	2468	2528	2588	2648	2708	2769	2829	2889	2949	3009
6	12	18	24	31	37	43	49	55	66	2506	2567	2628	2689	2750	2811	2872	2933	2994	3056
6	12	19	25	31	37	43	50	56	67	2544	2606	2668	2730	2792	2854	2916	2978	3040	3102
6	13	19	25	31	38	44	50	57	68	2581	2644	2707	2770	2833	2896	2959	3022	3085	3148
6	13	19	26	32	38	45	51	58	69	2619	2683	2747	2811	2875	2939	3003	3067	3131	3194
6	13	19	26	32	39	45	52	58	70	2657	2722	2787	2852	2917	2981	3046	3111	3176	3241
7	13	20	26	33	39	46	53	59	71	2695	2761	2827	2893	2958	3024	3090	3156	3221	3287
7	13	20	27	33	40	47	54	60	72	2733	2800	2867	2933	3000	3067	3133	3200	3267	3333
7	14	20	27	34	41	47	54	61	73	2771	2839	2906	2974	3042	3109	3177	3244	3312	3380
7	14	21	27	34	41	48	55	62	74	2809	2878	2946	3015	3083	3152	3220	3289	3357	3426
7	14	21	28	35	42	49	56	63	75	2847	2917	2986	3056	3125	3194	3264	3333	3403	3472
7	14	21	28	35	42	49	56	63	76	2885	2956	3026	3096	3167	3237	3307	3378	3448	3519
7	14	21	29	36	43	50	57	64	77	2923	2994	3066	3137	3208	3280	3351	3422	3494	3565
7	14	22	29	36	43	51	58	65	78	2961	3033	3106	3178	3250	3322	3394	3467	3539	3611
7	15	22	29	37	44	51	59	66	79	2999	3072	3145	3219	3292	3365	3438	3511	3584	3657
7	15	22	30	37	44	52	59	67	80	3037	3111	3185	3259	3333	3407	3481	3556	3630	3704
8	15	23	30	37	45	53	60	68	81	3075	3150	3225	3300	3375	3450	3525	3600	3675	3750
8	15	23	30	38	46	53	61	68	82	3113	3189	3265	3341	3417	3493	3569	3644	3720	3796
8	15	23	31	38	46	54	61	69	83	3151	3228	3305	3381	3458	3535	3612	3689	3766	3843
8	16	23	31	39	47	54	62	70	84	3189	3267	3344	3422	3500	3578	3656	3733	3811	3889
8	16	24	31	39	47	55	63	71	85	3227	3306	3384	3463	3542	3620	3699	3778	3856	3935
8	16	24	32	40	48	56	64	72	86	3265	3344	3424	3504	3583	3663	3743	3822	3902	3981
8	16	24	32	40	48	56	64	73	87	3303	3383	3464	3544	3625	3706	3786	3867	3947	4028
8	16	24	33	41	49	57	65	73	88	3341	3422	3504	3585	3667	3748	3830	3911	3993	4074
8	16	25	33	41	49	58	66	74	89	3379	3461	3544	3626	3708	3791	3873	3956	4038	4120
8	17	25	33	42	50	58	67	75	90	3417	3500	3583	3667	3750	3833	3917	4000	4083	4167
8	17	25	34	42	51	59	67	76	91	3455	3539	3623	3707	3792	3876	3960	4044	4129	4213
9	17	26	34	43	51	60	68	77	92	3493	3578	3663	3748	3833	3919	4004	4089	4174	4259
9	17	26	34	43	52	60	69	78	93	3531	3617	3703	3789	3875	3961	4047	4133	4219	4306
9	17	26	35	44	52	61	70	78	94	3569	3656	3743	3830	3917	4004	4091	4178	4265	4352
9	18	26	35	44	53	62	70	79	95	3606	3694	3782	3870	3958	4046	4134	4222	4310	4398
9	18	27	36	44	53	62	71	80	96	3644	3733	3822	3911	4000	4089	4178	4267	4356	4444
9	18	27	36	45	54	63	72	81	97	3682	3772	3862	3952	4042	4131	4221	4311	4401	4491
9	18	27	36	45	54	64	73	82	98	3720	3811	3902	3993	4083	4174	4265	4356	4446	4537
9	18	28	37	46	55	64	73	83	99	3758	3850	3942	4033	4125	4217	4308	4400	4492	4583
9	19	28	37	46	56	65	74	83	100	3796	3889	3981	4074	4167	4259	4352	4444	4537	4630
.1	.2	.3	.4	.5	.6	.7	.8	.9		41	42	43	44	45	46	47	48	49	50

Correction for tenths of height.

Correction for tenths of width.

	41	42	43	44	45	46	47	48	49	50
0.1	4	4	4	4	4	4	4	4	5	5
0.2	8	8	8	8	8	9	9	9	9	9
0.3	11	12	12	12	13	13	13	13	14	14
0.4	15	16	16	16	17	17	17	18	18	19
0.5	19	19	20	20	20	21	21	22	23	23
0.6	23	23	24	24	25	26	26	27	27	28
0.7	27	27	28	29	29	30	30	31	32	32
0.8	30	31	32	33	33	34	34	35	36	37
0.9	34	35	36	36	37	38	38	40	41	42

TABLE

— OF —

Prismoidal Corrections for Triangular Prismoids for Railway Earthwork,

IN CUBIC YARDS,

To be applied to the approximate volume found
by averaging end areas.

Correction for tenths of height—height′.

.1	.2	.3	.4	.5	.6	.7	.8	.9	w′−w	1	2	3	4	5	6	7	8	9	10
									1		1	1	1	2	2	2	2	3	3
								1	2	1	1	2	2	3	4	4	5	6	6
					1	1	1	1	3	1	2	3	4	5	6	6	7	8	9
				1	1	1	1	1	4	1	2	4	5	6	7	9	10	11	12
			1	1	1	1	1	1	5	2	3	5	6	8	9	11	12	14	15
		1	1	1	1	1	1	2	6	2	4	6	7	9	11	13	15	17	19
		1	1	1	1	2	2	2	7	2	4	6	9	11	13	15	17	19	22
		1	1	1	1	2	2	2	8	2	5	7	10	12	15	17	20	22	25
	1	1	1	1	2	2	2	3	9	3	6	8	11	14	17	19	22	25	28
	1	1	1	2	2	2	2	3	10	3	6	9	12	15	19	22	25	28	31
	1	1	1	2	2	2	3	3	11	3	7	10	14	17	20	24	27	31	34
	1	1	1	2	2	3	3	3	12	4	7	11	15	19	22	26	30	33	37
	1	1	2	2	2	3	3	4	13	4	8	12	16	20	24	28	32	36	40
	1	1	2	2	3	3	3	4	14	4	9	13	17	22	26	30	35	39	43
	1	1	2	2	3	3	4	4	15	5	9	14	19	23	28	32	37	42	46
	1	1	2	2	3	3	4	4	16	5	10	15	20	25	30	35	40	44	49
1	1	2	2	3	3	4	4	5	17	5	10	16	21	26	31	37	42	47	52
1	1	2	2	3	3	4	4	5	18	6	11	17	22	28	33	39	44	50	56
1	1	2	2	3	4	4	5	5	19	6	12	18	23	29	35	41	47	53	59
1	1	2	2	3	4	4	5	6	20	6	12	19	25	31	37	43	49	56	62
1	1	2	3	3	4	5	5	6	21	6	13	19	26	32	39	45	52	58	65
1	1	2	3	3	4	5	5	6	22	7	14	20	27	34	41	48	54	61	68
1	1	2	3	4	4	5	6	6	23	7	14	21	28	35	43	50	57	64	71
1	1	2	3	4	4	5	6	7	24	7	15	22	30	37	44	52	59	67	74
1	2	2	3	4	5	5	6	7	25	8	15	23	31	39	46	54	62	69	77
1	2	2	3	4	5	6	6	7	26	8	16	24	32	40	48	56	64	72	80
1	2	3	3	4	5	6	7	8	27	8	17	25	33	42	50	58	67	75	83
1	2	3	3	4	5	6	7	8	28	9	17	26	35	43	52	60	69	78	86
1	2	3	4	4	5	6	7	8	29	9	18	27	36	45	54	63	72	81	90
1	2	3	4	5	6	6	7	8	30	9	19	28	37	46	56	65	74	83	93
1	2	3	4	5	6	7	8	9	31	10	19	29	38	48	57	67	77	86	96
1	2	3	4	5	6	7	8	9	32	10	20	30	40	49	59	69	79	89	99
1	2	3	4	5	6	7	8	9	33	10	20	31	41	51	61	71	81	92	102
1	2	3	4	5	6	7	8	9	34	10	21	31	42	52	63	73	84	94	105
1	2	3	4	5	6	8	9	10	35	11	22	32	43	54	65	76	86	97	108
1	2	3	4	6	7	8	9	10	36	11	22	33	44	56	67	78	89	100	111
1	2	3	5	6	7	8	9	10	37	11	23	34	46	57	69	80	91	103	114
1	2	4	5	6	7	8	9	11	38	12	23	35	47	59	70	82	94	106	117
1	2	4	5	6	7	8	10	11	39	12	24	36	48	60	72	84	96	108	120
1	2	4	5	6	7	9	10	11	40	12	25	37	49	62	74	86	99	111	123
1	3	4	5	6	8	9	10	11	41	13	25	38	51	63	76	89	101	114	127
1	3	4	5	6	8	9	10	12	42	13	26	39	52	65	78	91	104	117	130
1	3	4	5	7	8	9	11	12	43	13	27	40	53	66	80	93	106	119	133
1	3	4	5	7	8	10	11	12	44	14	27	41	54	68	81	95	109	122	136
1	3	4	6	7	8	10	11	13	45	14	28	42	56	69	83	97	111	125	139
1	3	4	6	7	9	10	11	13	46	14	28	43	57	71	85	99	114	128	142
1	3	4	6	7	9	10	12	13	47	15	29	44	58	73	87	102	116	131	145
1	3	4	6	7	9	10	12	13	48	15	30	44	59	74	89	104	119	133	148
2	3	5	6	8	9	11	12	14	49	15	30	45	60	76	91	106	121	136	151
2	3	5	6	8	9	11	12	14	50	15	31	46	62	77	93	108	123	139	154

.1	.2	.3	.4	.5	.6	.7	.8	.9		1	2	3	4	5	6	7	8	9	10

Correction for tenths of width′— width.

	1	2	3	4	5	6	7	8	9	10
0.1									1	1
0.2									1	1
0.3						1	1	1	1	1
0.4					1	1	1	1	1	1
0.5				1	1	1	1	1	1	2
0.6			1	1	1	1	1	2	2	2
0.7			1	1	1	1	2	2	2	2
0.8			1	1	1	1	2	2	2	2
0.9		1	1	1	2	2	2	2	2	3

Correction for tenths of height'—height'.

.1	.2	.3	.4	.5	.6	.7	.8	.9		11	12	13	14	15	16	17	18	19	20
		1	1	1	1	2	2	2	7	24	26	28	30	32	35	37	39	41	43
		1	1	1	1	2	2	2	8	27	30	32	35	37	40	42	44	47	49
	1	1	1	1	2	2	2	2	9	31	33	36	39	42	44	47	50	53	56
	1	1	1	2	2	2	2	3	10	34	37	40	43	46	49	52	56	59	62
	1	1	1	2	2	2	3	3	11	37	41	44	48	51	54	58	61	65	68
	1	1	1	2	2	3	3	3	12	41	44	48	52	56	59	63	67	70	74
	1	1	2	2	2	3	3	4	13	44	48	52	56	60	64	68	72	76	80
	1	1	2	2	3	3	3	4	14	48	52	56	60	65	69	73	78	82	86
	1	1	2	2	3	3	4	4	15	51	56	60	65	69	74	79	83	88	93
	1	1	2	3	3	3	4	4	16	54	59	64	69	74	79	84	89	94	99
1	1	2	2	3	4	4	4	5	17	58	63	68	73	79	84	89	94	100	105
1	1	2	2	3	3	4	4	5	18	61	67	72	78	83	89	94	100	106	111
1	1	2	2	3	4	4	5	5	19	65	70	76	82	88	94	100	106	111	117
1	1	2	2	3	4	4	5	6	20	68	74	80	86	93	99	105	111	117	123
1	1	2	3	3	4	5	5	6	21	71	78	84	91	97	104	110	117	123	130
1	1	2	3	3	4	5	5	6	22	75	81	88	95	102	109	115	122	129	136
1	1	2	3	4	4	5	6	6	23	78	85	92	99	106	114	121	128	135	142
1	1	2	3	4	4	5	6	7	24	81	89	96	104	111	119	126	133	141	148
1	2	2	3	4	5	5	6	7	25	85	93	100	108	116	123	131	139	147	154
1	2	2	3	4	5	6	6	7	26	88	96	104	112	120	128	136	144	152	160
1	2	3	3	4	5	6	7	8	27	92	100	108	117	125	133	142	150	158	167
1	2	3	3	4	5	6	7	8	28	95	104	112	121	130	138	147	156	164	173
1	2	3	4	4	5	6	7	8	29	98	107	116	125	134	143	152	161	170	179
1	2	3	4	5	6	6	7	8	30	102	111	120	130	139	148	157	167	176	185
1	2	3	4	5	6	7	8	9	31	105	115	124	134	144	153	163	172	182	191
1	2	3	4	5	6	7	8	9	32	109	119	128	138	148	158	168	178	188	198
1	2	3	4	5	6	7	8	9	33	112	122	132	143	153	163	173	183	194	204
1	2	3	4	5	6	7	8	9	34	115	126	136	147	157	168	178	189	199	210
1	2	3	4	5	6	8	9	10	35	119	130	140	151	162	173	184	194	205	216
1	2	3	4	6	7	8	9	10	36	122	133	144	156	167	178	189	200	211	222
1	2	3	5	6	7	8	9	10	37	126	137	148	160	171	183	194	206	217	228
1	2	4	5	6	7	8	9	11	38	129	141	152	164	176	188	199	211	223	235
1	2	4	5	6	7	8	10	11	39	132	144	156	169	181	193	205	217	229	241
1	2	4	5	6	7	9	10	11	40	136	148	160	173	185	198	210	222	235	247
1	3	4	5	6	8	9	10	11	41	139	152	165	177	190	202	215	228	240	253
1	3	4	5	6	8	9	10	12	42	143	156	169	181	194	207	220	233	246	259
1	3	4	5	7	8	9	11	12	43	146	159	173	186	199	212	226	239	252	265
1	3	4	5	7	8	10	11	12	44	149	163	177	190	204	217	231	244	258	272
1	3	4	6	7	8	10	11	13	45	153	167	181	194	208	222	236	250	264	278
1	3	4	6	7	9	10	11	13	46	156	170	185	199	213	227	241	256	270	284
1	3	4	6	7	9	10	12	13	47	160	174	189	203	218	232	247	261	276	290
1	3	4	6	7	9	10	12	13	48	163	178	193	207	222	237	252	267	281	296
2	3	5	6	8	9	11	12	14	49	166	181	197	212	227	242	257	272	287	302
2	3	5	6	8	9	11	12	14	50	170	185	201	216	231	247	262	278	293	309
.1	.2	.3	.4	.5	.6	.7	.8	.9		11	12	13	14	15	16	17	18	19	20

Correction for tenths of width'—width.

	11	12	13	14	15	16	17	18	19	20
0.1							1	1	1	1
0.2	1	1	1	1	1	1	1	1	1	1
0.3	1	1	1	1	1	1	2	2	2	2
0.4	1	1	2	2	2	2	2	2	2	2
0.5	2	2	2	2	2	2	2	3	3	3
0.6	2	2	2	3	3	3	3	3	4	4
0.7	2	3	3	3	3	3	3	4	4	4
0.8	3	3	3	3	4	4	4	4	5	5
0.9	3	3	3	4	4	4	4	5	5	6

Correction for tenths of height'—height.

.1	.2	.3	.4	.5	.6	.7	.8	.9	w'−w.	21	22	23	24	25	26	27	28	29	30
								1	1	6	7	7	7	8	8	8	9	9	9
							1	1	2	13	14	14	15	15	16	17	17	18	19
					1	1	1	1	3	19	20	21	22	23	24	25	26	27	28
				1	1	1	1	1	4	26	27	28	30	31	32	33	35	36	37
			1	1	1	1	1	1	5	32	34	35	37	39	40	42	43	45	46
		1	1	1	1	1	1	2	6	39	41	43	44	46	48	50	52	54	56
		1	1	1	1	2	2	2	7	45	48	50	52	54	56	58	60	63	65
		1	1	1	1	2	2	2	8	52	54	57	59	62	64	67	69	72	74
	1	1	1	1	2	2	2	3	9	58	61	64	67	69	72	75	78	81	83
	1	1	1	2	2	2	2	3	10	65	68	71	74	77	80	83	86	90	93
	1	1	1	2	2	2	3	3	11	71	75	78	81	85	88	92	95	98	102
	1	1	1	2	2	3	3	3	12	78	81	85	89	93	96	100	104	107	111
	1	1	2	2	2	3	3	4	13	84	88	92	96	100	104	108	112	116	120
	1	1	2	2	3	3	3	4	14	91	95	99	104	108	112	117	121	125	130
	1	1	2	2	3	3	4	4	15	97	102	106	111	116	120	125	130	134	139
	1	1	2	2	3	3	4	4	16	104	109	114	119	123	128	133	138	143	148
1	1	2	2	3	3	4	4	5	17	110	115	121	126	131	136	142	147	152	157
1	1	2	2	3	3	4	4	5	18	117	122	128	133	139	144	150	156	161	167
1	1	2	2	3	4	4	5	5	19	123	129	135	141	147	152	158	164	170	176
1	1	2	2	3	4	4	5	6	20	130	136	142	148	154	160	167	173	179	185
1	1	2	3	3	4	5	5	6	21	136	143	149	156	162	169	175	181	188	194
1	1	2	3	3	4	5	5	6	22	143	149	156	163	170	177	183	190	197	204
1	1	2	3	4	4	5	6	6	23	149	156	163	170	177	185	192	199	206	213
1	1	2	3	4	4	5	6	7	24	156	163	170	178	185	193	200	207	215	222
1	2	2	3	4	5	5	6	7	25	162	170	177	185	193	201	208	216	224	231
1	2	2	3	4	5	6	6	7	26	169	177	185	193	201	209	217	225	233	241
1	2	3	3	4	5	6	7	8	27	175	183	192	200	208	217	225	233	242	250
1	2	3	3	4	5	6	7	8	28	181	190	199	207	216	225	233	242	251	259
1	2	3	4	4	5	6	7	8	29	188	197	206	215	224	233	242	251	260	269
1	2	3	4	5	6	6	7	8	30	194	204	213	222	231	241	250	259	269	278
1	2	3	4	5	6	7	8	9	31	201	210	220	230	239	249	258	268	277	287
1	2	3	4	5	6	7	8	9	32	207	217	227	237	247	257	267	277	286	296
1	2	3	4	5	6	7	8	9	33	214	224	234	244	255	265	275	285	295	306
1	2	3	4	5	6	7	8	9	34	220	231	241	252	262	273	283	294	304	315
1	2	3	4	5	6	8	9	10	35	227	238	248	259	270	281	292	302	313	324
1	2	3	4	6	7	8	9	10	36	233	244	256	267	278	289	300	311	322	333
1	2	3	5	6	7	8	9	10	37	240	251	263	274	285	297	308	320	331	343
1	2	4	5	6	7	8	10	11	38	246	258	270	281	293	305	317	328	340	352
1	2	4	5	6	7	9	10	11	39	253	265	277	289	301	313	325	337	349	361
1	2	4	5	6	7	9	10	11	40	259	272	284	296	309	321	333	346	358	370
1	3	4	5	6	8	9	10	11	41	266	278	291	304	316	329	342	354	367	380
1	3	4	5	6	8	9	11	12	42	272	285	298	311	324	337	350	363	376	389
1	3	4	5	7	8	10	11	12	43	279	292	305	319	332	345	358	372	385	398
1	3	4	5	7	8	10	11	12	44	285	299	312	326	340	353	367	380	394	407
1	3	4	6	7	8	10	11	13	45	292	306	319	333	347	361	375	389	403	417
1	3	4	6	7	9	10	11	13	46	298	312	327	341	355	369	383	398	412	426
1	3	4	6	7	9	10	12	13	47	305	319	334	348	363	377	392	406	421	435
1	3	4	6	7	9	10	12	13	48	311	326	341	356	370	385	400	415	430	444
2	3	5	6	8	9	11	12	14	49	318	333	348	363	378	393	408	423	439	454
2	3	5	6	8	9	11	12	14	50	324	340	355	370	386	401	417	432	448	463

.1	.2	.3	.4	.5	.6	.7	.8	.9		21	22	23	24	25	26	27	28	29	30

Correction for tenths of width'— width.

	21	22	23	24	25	26	27	28	29	30
0.1	1	1	1	1	1	1	1	1	1	1
0.2	1	1	1	1	2	2	2	2	2	2
0.3	2	2	2	2	2	2	3	3	3	3
0.4	3	3	3	3	3	3	3	3	4	4
0.5	3	3	4	4	4	4	4	4	4	5
0.6	4	4	4	4	5	5	5	5	6	6
0.7	5	5	5	5	5	6	6	6	6	6
0.8	5	5	6	6	6	6	7	7	7	7
0.9	6	6	6	6	7	7	7	8	8	8

Correction for tenths of height—height.

.1	.2	.3	.4	.5	.6	.7	.8	.9	w'—w.	31	32	33	34	35	36	37	38	39	40
									1	10	10	10	10	11	11	11	12	12	12
								1	2	19	20	20	21	22	22	23	23	24	25
					1	1	1	1	3	29	30	31	31	32	33	34	35	36	37
				1	1	1	1	1	4	38	40	41	42	43	44	46	47	48	49
			1	1	1	1	1	1	5	48	49	51	52	54	56	57	59	60	62
		1	1	1	1	1	1	2	6	57	59	61	63	65	67	69	70	72	74
	1	1	1	1	1	2	2	2	7	67	69	71	73	76	78	80	82	84	86
	1	1	1	1	1	2	2	2	8	77	79	81	84	86	89	91	94	96	99
1	1	1	2	2	2	2	2	2	9	86	89	92	94	97	100	103	106	108	111
1	1	1	2	2	2	2	2	3	10	96	99	102	105	108	111	114	117	120	123
1	1	1	2	2	2	3	3	3	11	105	109	112	115	119	122	126	129	132	136
1	1	1	2	2	3	3	3	4	12	115	119	122	126	130	133	137	141	144	148
1	1	2	2	2	3	3	3	4	13	124	128	132	136	140	144	148	152	156	160
1	1	2	2	3	3	3	4	4	14	134	138	143	147	151	156	160	164	169	173
1	1	2	2	3	3	4	4	4	15	144	148	153	157	162	167	171	176	181	185
	1	2	2	3	3	4	4	4	16	153	158	163	168	173	178	183	188	193	198
1	1	2	2	3	3	4	4	5	17	163	168	173	178	184	189	194	199	205	210
1	1	2	2	3	3	4	4	5	18	172	178	183	189	194	200	206	211	217	222
1	1	2	2	3	4	4	5	5	19	182	188	194	199	205	211	217	223	229	235
1	1	2	2	3	4	4	5	6	20	191	198	204	210	216	222	228	235	241	247
1	1	2	3	3	4	5	5	6	21	201	207	214	220	227	233	240	246	253	259
1	1	2	3	3	4	5	5	6	22	210	217	224	231	238	244	251	258	265	272
1	1	2	3	4	4	5	6	6	23	220	227	234	241	248	256	263	270	277	284
1	1	2	3	4	4	5	6	7	24	230	237	244	252	259	267	274	281	289	296
1	2	2	3	4	5	5	6	7	25	239	247	255	262	270	278	285	293	301	309
1	2	2	3	4	5	6	6	7	26	249	257	265	273	281	289	297	305	313	321
1	2	3	3	4	5	6	7	8	27	258	267	275	283	292	300	308	317	325	333
1	2	3	4	4	5	6	7	8	28	268	277	285	294	302	311	320	328	337	346
1	2	3	4	4	5	6	7	8	29	277	286	295	304	313	322	331	340	349	358
1	2	3	4	5	6	6	7	8	30	287	296	306	315	324	333	343	352	361	370
1	2	3	4	5	6	7	8	9	31	297	306	316	325	335	344	354	364	373	383
1	2	3	4	5	6	7	8	9	32	306	316	326	336	346	356	365	375	385	395
1	2	3	4	5	6	7	8	9	33	316	326	336	346	356	367	377	387	397	407
1	2	3	4	5	6	7	8	9	34	325	336	346	357	367	378	388	399	409	420
1	2	3	4	5	6	8	9	10	35	335	346	356	367	378	389	400	410	421	432
1	2	3	4	6	7	8	9	10	36	344	356	367	378	389	400	411	422	433	444
1	2	3	5	6	7	8	9	10	37	354	365	377	388	400	411	423	434	445	457
1	2	4	5	6	7	8	10	11	38	364	375	387	399	410	422	434	446	457	469
1	2	4	5	6	7	8	10	11	39	373	385	397	409	421	433	445	457	469	481
1	2	4	5	6	7	9	10	11	40	383	395	407	420	432	444	457	469	481	494
1	3	4	5	6	8	9	10	11	41	392	405	418	430	443	456	468	481	494	506
1	3	4	5	6	8	9	10	12	42	402	415	428	441	454	467	480	493	506	519
1	3	4	5	7	8	9	11	12	43	411	425	438	451	465	478	491	504	518	531
1	3	4	5	7	8	10	11	12	44	421	435	448	462	475	489	502	516	530	543
1	3	4	6	7	8	10	11	13	45	431	444	458	472	486	500	514	528	542	556
1	3	4	6	7	9	10	11	13	46	440	454	469	483	497	511	525	540	554	568
1	3	4	6	7	9	10	12	13	47	450	464	479	493	508	522	537	551	566	580
1	3	4	6	7	9	10	12	13	48	459	474	489	504	519	533	548	563	578	593
2	3	5	6	8	9	11	12	14	49	469	484	499	514	529	544	560	575	590	605
2	3	5	6	8	9	11	12	14	50	478	494	509	525	540	556	571	586	602	617

.1	.2	.3	.4	.5	.6	.7	.8	.9		31	32	33	34	35	36	37	38	39	40
									0.1	1	1	1	1	1	1	1	1	1	1
									0.2	2	2	2	2	2	2	2	2	2	2
									0.3	3	3	3	3	3	3	3	4	4	4
Correction for tenths									0.4	4	4	4	4	4	4	5	5	5	5
									0.5	5	5	5	5	5	5	6	6	6	6
of width'—width.									0.6	6	6	6	6	6	6	7	7	7	7
									0.7	7	7	7	7	8	8	8	8	8	9
									0.8	8	8	8	8	9	9	9	9	10	10
									0.9	9	9	9	9	10	10	10	11	11	11

Correction for tenths of height—height′.

.1	.2	.3	.4	.5	.6	.7	.8	.9	w′—w.	41	42	43	44	45	46	47	48	49	50
								I	1	13	13	13	14	14	14	15	15	15	15
							I	I	2	25	26	27	27	28	28	29	30	30	31
					I	I	I	I	3	38	39	40	41	42	43	44	44	45	46
				I	I	I	I	I	4	51	52	53	54	56	57	58	59	60	62
			I	I	I	I	I	I	5	63	65	66	68	69	71	73	74	76	77
		I	I	I	I	I	I	2	6	76	78	80	81	83	85	87	89	91	93
		I	I	I	I	2	2	2	7	89	91	93	95	97	99	102	104	106	108
		I	I	I	I	2	2	2	8	101	104	106	109	111	114	116	119	121	123
	I	I	I	I	2	2	2	2	9	114	117	119	122	125	128	131	133	136	139
	I	I	I	2	2	2	2	3	10	127	130	133	136	139	142	145	148	151	154
	I	I	I	2	2	2	3	3	11	139	143	146	149	153	156	160	163	166	170
	I	I	I	2	2	3	3	3	12	152	156	159	163	167	170	174	178	181	185
	I	I	2	2	2	3	3	4	13	165	169	173	177	181	185	189	193	197	201
	I	I	2	2	3	3	4	4	14	177	181	186	190	194	199	203	207	212	216
	I	I	2	2	3	3	4	4	15	190	194	199	204	208	213	218	222	227	231
	I	2	3	3	3	4	4	4	16	202	207	212	217	222	227	232	237	242	247
	I	2	3	3	4	4	4	5	17	215	220	226	231	236	241	247	252	257	262
I	I	2	3	3	4	4	5	5	18	228	233	239	244	250	256	261	267	272	278
I	I	2	3	4	4	5	5	5	19	240	246	252	258	264	270	276	281	287	293
I	I	2	3	4	4	5	5	6	20	253	259	265	272	278	284	290	296	302	309
I	I	2	3	3	4	5	5	6	21	266	272	279	285	292	298	305	311	318	324
I	I	2	3	4	5	5	6	6	22	278	285	292	299	306	312	319	326	333	340
I	I	2	3	4	4	5	6	6	23	291	298	305	312	319	327	334	341	348	355
I	I	2	3	4	4	5	6	7	24	304	311	319	326	333	341	348	356	363	370
I	2	2	3	4	5	5	6	7	25	316	324	332	340	347	355	363	370	378	386
I	2	2	3	4	5	6	6	7	26	329	337	345	353	361	369	377	385	393	401
I	2	3	3	4	5	6	7	8	27	342	350	358	367	375	383	392	400	408	417
I	2	3	3	4	5	6	7	8	28	354	363	372	380	389	398	406	415	423	432
I	2	3	4	4	5	6	7	8	29	367	376	385	394	403	412	421	430	439	448
I	2	3	4	5	6	6	7	8	30	380	389	398	407	417	426	435	444	454	463
I	2	3	4	5	6	7	8	9	31	392	402	411	421	431	440	450	459	469	478
I	2	3	4	5	6	7	8	9	32	405	415	425	435	444	454	464	474	484	494
I	2	3	4	5	6	7	8	9	33	418	428	438	448	458	469	479	489	499	509
I	2	3	4	5	6	7	8	9	34	430	441	451	462	472	483	493	504	514	525
I	2	3	4	5	6	8	9	10	35	443	454	465	475	486	497	508	519	529	540
I	2	3	4	6	7	8	9	10	36	456	467	478	489	500	511	522	533	544	556
I	2	3	5	6	7	8	9	10	37	468	480	491	502	514	525	537	548	560	571
I	2	4	5	6	7	8	9	11	38	481	493	504	516	528	540	551	563	575	586
I	2	4	5	6	7	8	10	11	39	494	506	518	530	542	554	566	578	590	602
I	2	4	5	6	7	9	10	11	40	506	519	531	543	556	568	580	593	605	617
I	3	4	5	6	8	9	10	11	41	519	531	544	557	569	582	595	607	620	633
I	3	4	5	6	8	9	10	12	42	531	544	557	570	583	596	609	622	635	648
I	3	4	5	7	8	9	11	12	43	544	557	571	584	597	610	624	637	650	664
I	3	4	5	7	8	10	11	12	44	557	570	584	598	611	625	638	652	665	679
I	3	4	6	7	8	10	11	13	45	569	583	597	611	625	639	653	667	681	694
I	3	4	6	7	9	10	11	13	46	582	596	610	625	639	653	667	681	696	710
I	3	4	6	7	9	10	12	13	47	595	609	624	638	653	667	682	696	711	725
I	3	4	6	7	9	10	12	13	48	607	622	637	652	667	681	696	711	726	741
2	3	5	6	8	9	11	12	14	49	620	635	650	665	681	696	711	726	741	756
2	3	5	6	8	9	11	12	14	50	633	648	664	679	694	710	725	741	756	772
.1	.2	.3	.4	.5	.6	.7	.8	.9		41	42	43	44	45	46	47	48	49	50

Correction for tenths		41	42	43	44	45	46	47	48	49	50
of width′— width.	0.1	1	1	1	1	1	1	1	1	2	2
	0.2	3	3	3	3	3	3	3	3	3	3
	0.3	4	4	4	4	4	4	4	4	5	5
	0.4	5	5	5	5	6	6	6	6	6	6
	0.5	6	6	7	7	7	7	7	7	8	8
	0.6	8	8	8	8	8	9	9	9	9	9
	0.7	9	9	9	10	10	10	10	10	11	11
	0.8	10	10	11	11	11	11	12	12	12	12
	0.9	11	12	12	12	13	13	13	13	14	14

www.ingramcontent.com/pod-product-compliance
Lightning Source LLC
Chambersburg PA
CBHW022030190326
41519CB00010B/1651